T0304636

# Advances in Polymer Derived Ceramics and Composites

# Advances in Polymer Derived Ceramics and Composites

Ceramic Transactions, Volume 213

*A Collection of Papers Presented at the 8th Pacific Rim Conference on Ceramic and Glass Technology*
*May 31–June 5, 2009*
*Vancouver, British Columbia*

Edited by
Paolo Colombo
Rishi Raj

Volume Editor
Mrityunjay Singh

A John Wiley & Sons, Inc., Publication

Copyright © 2010 by The American Ceramic Society. All rights reserved.

Published by John Wiley & Sons, Inc., Hoboken, New Jersey.
Published simultaneously in Canada.

No part of this publication may be reproduced, stored in a retrieval system, or transmitted in any form
or by any means, electronic, mechanical, photocopying, recording, scanning, or otherwise, except as
permitted under Section 107 or 108 of the 1976 United States Copyright Act, without either the prior
written permission of the Publisher, or authorization through payment of the appropriate per-copy fee to
the Copyright Clearance Center, Inc., 222 Rosewood Drive, Danvers, MA 01923, (978) 750-8400, fax
(978) 750-4470, or on the web at www.copyright.com. Requests to the Publisher for permission should
be addressed to the Permissions Department, John Wiley & Sons, Inc., 111 River Street, Hoboken, NJ
07030, (201) 748-6011, fax (201) 748-6008, or online at http://www.wiley.com/go/permission.

Limit of Liability/Disclaimer of Warranty: While the publisher and author have used their best efforts in
preparing this book, they make no representations or warranties with respect to the accuracy or
completeness of the contents of this book and specifically disclaim any implied warranties of
merchantability or fitness for a particular purpose. No warranty may be created or extended by sales
representatives or written sales materials. The advice and strategies contained herein may not be
suitable for your situation. You should consult with a professional where appropriate. Neither the
publisher nor author shall be liable for any loss of profit or any other commercial damages, including
but not limited to special, incidental, consequential, or other damages.

For general information on our other products and services or for technical support, please contact our
Customer Care Department within the United States at (800) 762-2974, outside the United States at
(317) 572-3993 or fax (317) 572-4002.

Wiley also publishes its books in a variety of electronic formats. Some content that appears in print may
not be available in electronic format. For information about Wiley products, visit our web site at
www.wiley.com.

*Library of Congress Cataloging-in-Publication Data is available.*

ISBN 978-0-470-87800-2

# Contents

## PROCESSING AND APPLICATIONS

# Preface

Polymer-Derived Ceramics (PDCs) are a new class of ceramic materials, which as their name suggests, are made by direct conversion of a polymer into the ceramic phase. Up to this point silicon based organics have been used to made materials constituted from Si, C, N and O. The ceramic remains amorphous up to very high temperatures, and apparently, retains a memory of its organic molecular network. The combination of chemical design of the precursors, and the range of processing variables hold the potential of creating molecularly engineered ceramics with novel multifunctional properties, which are not accessible in conventional ceramics. These properties include not only remarkable thermal stability, high oxidation and thermal shock resistance, and zero creep behavior, that are important in structural applications, but also functional properties such semiconductivity (up to at least 1300°C), photoluminescence, and piezoresistivity. These properties appear to derive from the nanoscale structure of the PDCs, which, in general, is amorphous but yields strong and persistent small–scale structure in Small Angle X-ray Scattering. This structure has been modeled as a three dimensional network of graphene sheets with a domain size of approximately 1 to 5 nm. These domains encompass Si-O, Si-C and Si-N bonds: their relative populations depending on the composition of the material.

The polymer-based processing route permits cost-effective routes for net–shape forming, which are commonly used for make shapes from organics. A wide range of components have been produced, including high-strength, high-temperature, ceramic fibers, ceramic matrix composites, ceramic coatings, porous bodies, highly selective membranes, and miniaturized devices (MEMS). The organic precursors have also been used as reactive binders for metal powders and for joining ceramic bodies.

The PDCs hold the potential of spawning a new area of research, one that forms a continuum between polymers and ceramics. The organic precursors can be converted partially into a ceramic, thus yielding a new class of hybrid materials that lie at the interface of organics and inorganics. The novelty of these structures would be their use at temperatures that are normally too high for organic polymers, in the 300°C to 600°C range. These materials can retain the low density of the organic phase, yet incorporate high chemical inertness and mechanical properties. The functional properties of such hybrid materials have yet to be explored in detail.

In recent years, a vibrant research community, that includes a large number of young scientists, has grown around the processing, structure and properties of PDCs. These scientists draw from various fields: chemistry, physics and engineering, providing the interdisciplinary foundation for research on these materials. Support from funding agencies in the US, Europe and Asia is growing rapidly.

This book collects some of papers presented at the very successful Symposium "Polymer Derived Ceramics and Composites" in the framework of the 8th Pacific Rim Conference on Ceramic and Glass Technology. There, over 70 researchers from around the world discussed their latest innovations over four full days. It covers all the main aspects of interdisciplinary research and development in the field of Polymer-Derived-Ceramics, from the precursor synthesis and characteristics to the polymer-to-ceramic conversion, from processing and shaping of preceramic polymers into ceramic components to their microstructure at the nano- and micro-scale, from their properties to their most relevant applications in different fields.

We hope that it will stimulate further activities in this fascinating field of new materials.

PAOLO COLOMBO
RISHI RAJ

# Introduction

The 8th Pacific Rim Conference on Ceramic and Glass Technology (PACRIM 8), was the eighth in a series of international conferences that provided a forum for presentations and information exchange on the latest emerging ceramic and glass technologies. The conference series began in 1993 and has been organized in USA, Korea, Japan, China, and Canada. PACRIM 8 was held in Vancouver, British Columbia, Canada, May 31–June 5, 2009 and was organized and sponsored by The American Ceramic Society. Over the years, PACRIM conferences have established a strong reputation for the state-of-the-art presentations and information exchange on the latest emerging ceramic and glass technologies. They have facilitated global dialogue and discussion with leading world experts.

The technical program of PACRIM 8 covered wide ranging topics and identified global challenges and opportunities for various ceramic technologies. The goal of the program was also to generate important discussion on where the particular field is heading on a global scale. It provided a forum for knowledge sharing and to make new contacts with peers from different continents.

The program also consisted of meetings of the International Commission on Glass (ICG), and the Glass and Optical Materials and Basic Science divisions of The American Ceramic Society. In addition, the International Fulrath Symposium on the role of new ceramic technologies for sustainable society was also held. The technical program consisted of more than 900 presentations from 41 different countries. A selected group of peer reviewed papers have been compiled into seven volumes of The American Ceramic Society's Ceramic Transactions series (Volumes 212-218) as outlined below:

- **Innovative Processing and Manufacturing of Advanced Ceramics and Composites, Ceramic Transactions, Vol. 212,** Zuhair Munir, Tatsuki Ohji, and Koji Watari, Editors; Mrityunjay Singh, Volume Editor
  *Topics in this volume include Synthesis and Processing by the Spark Plasma*

*Method; Novel, Green, and Strategic Processing; and Advanced Powder Processing*

- **Advances in Polymer Derived Ceramics and Composites, Ceramic Transactions, Vol. 213,** Paolo Colombo and Rishi Raj, Editors; Mrityunjay Singh, Volume Editor
  *This volume includes papers on polymer derived fibers, composites, functionally graded materials, coatings, nanowires, porous components, membranes, and more.*

- **Nanostructured Materials and Systems, Ceramic Transactions, Vol. 214,** Sanjay Mathur and Hao Shen, Editors; Mrityunjay Singh, Volume Editor
  *Includes papers on the latest developments related to synthesis, processing and manufacturing technologies of nanoscale materials and systems including one-dimensional nanostructures, nanoparticle-based composites, electrospinning of nanofibers, functional thin films, ceramic membranes, bioactive materials and self-assembled functional nanostructures and nanodevices.*

- **Design, Development, and Applications of Engineering Ceramics and Composite Systems, Ceramic Transactions, Vol. 215, Dileep Singh, Dongming Zhu, and Yanchum Zhou; Mrityunjay Singh, Volume Editor**
  *Includes papers on design, processing and application of a wide variety of materials ranging from SiC SiAlON, $ZrO_2$, fiber reinforced composites; thermal/environmental barrier coatings; functionally gradient materials; and geopolymers.*

- **Advances in Multifunctional Materials and Systems, Ceramic Transactions, Vol. 216,** Jun Akedo, Hitoshi Ohsato, and Takeshi Shimada, Editors; Mrityunjay Singh, Volume Editor
  *Topics dealing with advanced electroceramics including multilayer capacitors; ferroelectric memory devices; ferrite circulators and isolators; varistors; piezoelectrics; and microwave dielectrics are included.*

- **Ceramics for Environmental and Energy Systems, Ceramic Transactions, Vol. 217,** Aldo Boccaccini, James Marra, Fatih Dogan, and Hua-Tay Lin, Editors; Mrityunjay Singh, Volume Editor
  *This volume includes selected papers from four symposia: Glasses and Ceramics for Nuclear and Hazardous Waste Treatment; Solid Oxide Fuel Cells and Hydrogen Technology; Ceramics for Electric Energy Generation, Storage, and Distribution; and Photocatalytic Materials.*

- **Advances in Bioceramics and Biotechnologies, Ceramic Transactions, Vol. 218;** Roger Narayan and Joanna McKittrick, Editors; Mrityunjay Singh, Volume Editor
  *Includes selected papers from two cutting edge symposia: Nano-Biotechnology and Ceramics in Biomedical Applications and Advances in Biomineralized Ceramics, Bioceramics, and Bioinspiried Designs.*

I would like to express my sincere thanks to Greg Geiger, Technical Content Manager of The American Ceramic Society for his hard work and tireless efforts in the publication of this series. I would also like to thank all the contributors, editors, and reviewers for their efforts.

MRITYUNJAY SINGH
Volume Editor and Chairman, PACRIM-8
Ohio Aerospace Institute
Cleveland, OH (USA)

# Synthesis and Characterization

POLY[(SILYLYNE)ETHYNYLENE] AND POLY[(SILYLENE)ETHYNYLENE]: NEW
PRECURSORS FOR THE EFFICIENT SYNTHESIS OF SILICON CARBIDE

Soichiro Kyushin,[a] Hiroyuki Shiraiwa,[a] Masafumi Kubota,[a] Keisuke Negishi,[a] and Kiyohito Okamura,[b]
and Kenji Suzuki[b]
a) Department of Chemistry and Chemical Biology, Graduate School of Engineering, Gunma
    University, Kiryu, Gunma 376-8515, Japan
b) Advanced Institute of Materials Science, Sendai, Miyagi 982-0252, Japan

ABSTRACT
        Poly[(silylyne)ethynylene] and poly[(silylene)ethynylene] were found to become new
precursors for the efficient synthesis of Si–C–O ceramics.   These precursors were synthesized by the
reactions of dilithioacetylene with trichlorosilane or dichlorosilane in one step.   When these
precursors were heated gradually up to 1000 °C, Si–C–O ceramics were formed in high yields.   The
weight loss is 5% in the case of poly[(silylyne)ethynylene] and 10% in the case of
poly[(silylene)ethynylene].   The remarkably low weight losses enable the molding of the Si–C–O
ceramics and the formation of thin films.

INTRODUCTION
        Silicon carbide and related ceramics have been used as materials which are thermally stable
and have high tensile strength.   Silicon carbide fibers have been industrially produced by the thermal
rearrangement of poly(dimethylsilane) into polycarbosilane and the thermolysis of polycarbosilane at
more than 1000 °C according to Yajima's method.[1,2]   This method was well established, and no other
methods could not replace it for more than thirty years.   However, in order to improve versatility and
applicability of silicon carbide and related ceramics, development of other synthetic routes is required
and has been studied.[3,4]
        Our concept for designing a novel synthetic route is based on the formation of a network
structure during the thermolysis of the precursor.   If a network structure was highly developed by the
formation of silicon–carbon bonds, the precursor might be expected to form silicon carbide efficiently.
As such silicon–carbon bond formation, we focused our attention on hydrosilylation of carbon–carbon
triple   bonds   with   hydrosilanes,   and   planned   to   use   poly-[(silylyne)ethynylene]   and
poly[(silylene)ethynylene] as precursors.   We report herein the synthesis of these precursors and their
thermolysis leading to the formation of Si–C–O ceramics without significant weight losses.

## SYNTHESIS OF POLY[(SILYLYNE)ETHYNYLENE] AND POLY[(SILYLENE)ETHYNYLENE]

Poly[(silylyne)ethynylene] (**1**) and poly[(silylene)ethynylene] (**2**) were synthesized by the reactions of dilithioacetylene, which was prepared from trichloroethylene and *n*-butyllithium,[5] with trichlorosilane or dichlorosilane in 82 and 42% yields, respectively. Compound **1** was obtained as pale yellow powder. Compound **2** was obtained initially as pale yellow viscous oil, but after the oil was fully dried under reduced pressure, this compound was obtained as pale yellow powder. The IR spectra of these compounds show the Si–H stretching band at 2190–2260 $cm^{-1}$ and the C≡C stretching band at 2050–2060 $cm^{-1}$. As the Si–O stretching band is also observed at 1070–1110 $cm^{-1}$, these compounds contain oxygen atoms.

## THERMOLYSIS OF POLY[(SILYLYNE)ETHYNYLENE] AND POLY[(SILYLENE)ETHYNYLENE]

When these precursors were gradually heated up to 1000 °C under a nitrogen atmosphere, black powder was obtained in 95% yield from **1** and in 90% yield from **2**. The weight losses of 5 and 10% are remarkably low compared with Yajima's method, where polycarbosilane loses about 40% of its weight on thermolysis to silicon carbide. The elemental analysis showed that the product from **1** contains silicon, carbon, and oxygen atoms in 39, 32, and 21%, respectively, and the product from **2** contains silicon, carbon, and oxygen atoms in 47, 22, and 20%, respectively. These data show the products are Si–C–O ceramics. The structures of the Si–C–O ceramics were analyzed by X-ray diffraction, but no reflections were observed, indicating that the Si–C–O ceramics are amorphous.

$$1 \xrightarrow{1000\ °C} \begin{matrix} Si\text{–}C\text{–}O \\ 95\% \end{matrix}$$

$$2 \xrightarrow{1000\ °C} \begin{matrix} Si\text{–}C\text{–}O \\ 90\% \end{matrix}$$

In order to obtain more information about thermal behavior of **1** and **2**, TG-DTA measurements were carried out (Figure 1). Exothermic peaks were observed at 266 °C in the case of **1** and at 222 °C in the case of **2**. These exothermic peaks are ascribed to hydrosilylation of the precursors. The weight losses of **1** and **2** at 1500 °C are 4.2 and 6.7%, respectively. Although these values are slightly smaller than the above values in the synthetic experiments, both data show very low weight losses. This thermal stability of the Si–C–O ceramics is unusual because silicon carbide containing oxygen atoms has been known to evolve SiO and CO and reduce the weight considerably on heating from 1000 °C to 1500 °C.[6] This unusual thermal stability of the Si–C–O ceramics cannot be explained clearly, but the development of the rigid network structure by hydrosilylation might be responsible for this thermal stability.

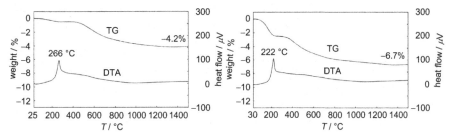

Figure 1.    TG-DTA data of **1** (left) and **2** (right) measured under a nitrogen atmosphere.

## MOLDING AND FORMATION OF A THIN FILM OF THE Si–C–O CERAMICS

The low weight losses enable the molding of the Si–C–O ceramics. As mentioned above, the powder of **1** gave black powder of the Si–C–O ceramics without a significant weight loss (Figure 2 (top)). On the other hand, a disk-shaped ceramic pellet was obtained by pressing the powder of precursor **1** into a disk-shaped pellet and heating it up to 1000 °C (Figure 2 (bottom)). Also, a thin film was obtained by spreading the oil of **2** on a quartz plate by spin coating technique and heating it up to 1000 °C. These methods expand the versatility and applicability of silicon carbide and related ceramics.

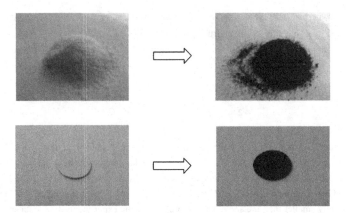

Figure 2. Molding of the Si–C–O ceramics by using precursor **1**.

CONCLUSION

In summary, we have found that poly[(silylyne)ethynylene] and poly[(silylene)ethynylene] can be used as precursors for the efficient synthesis of Si–C–O ceramics. The weight losses on heating the precursors up to 1500 °C are remarkably low probably due to the formation of a rigid network structure by hydrosilylation. These precursors provide the Si–C–O ceramics as powder, molds, and thin films.

REFERENCES

[1]S. Yajima, J. Hayashi, and M. Omori, *Chem. Lett.*, 931 (1975).

[2]S. Yajima, K. Okamura, J. Hayashi, and M. Omori, *J. Am. Ceram. Soc.*, **59**, 324 (1976).

[3]L. V. Interrante, K. Moraes, Q. Liu, N. Lu, A. Puerta, and L. G. Sneddon, *Pure Appl. Chem.*, **74**, 2111 (2002).

[4]J. Clade, E. Seider, and D. Sporn, *J. Eur. Ceram. Soc.*, **25**, 123 (2005).

[5]S. Ijadi-Maghsoodi, Y. Peng, and T. J. Barton, *J. Polym. Sci. A, Polym. Chem.*, **28**, 955 (1990).

[6]T. Shimoo, H. Chen, and K. Okamura, *J. Ceram. Soc. Jpn.*, **100**, 48 (1992).

# SYNTHESIS OF A CATALYST-LOADED SiC MATERIAL FROM Si-BASED POLYMER

Akira Idesaki, Masaki Sugimoto, Masahito Yoshikawa
Quantum Beam Science Directorate, Japan Atomic Energy Agency
Takasaki, Gunma, JAPAN

ABSTRACT

A novel precursor composed of polycarbosilane (PCS) and palladium(II) acetate (Pd(OAc)$_2$) was synthesized and characterized in order to develop a catalyst-loaded SiC material. It was found that PCS and Pd(OAc)$_2$ react with each other through crosslinking reaction when their solution are mixed and that the mass residue of synthesized precursor at 1473K is 70% which is 14% higher than that of PCS. The catalytic performance of Pd-loaded SiC material obtained by the pyrolysis of precursor was evaluated by evolution of CO$_2$ gas due to oxidation of CO gas. It was found that the Pd-loaded SiC material shows a catalytic performance.

INTRODUCTION

In recent years, increasing attention has been paid to global warming caused by greenhouse gases such as carbon dioxide (CO$_2$) and methane (CH$_4$). In order to reduce and utilize the greenhouse gases, the reaction of CO$_2$ reforming of CH$_4$ (CO$_2$ + CH$_4$ → 2CO + 2H$_2$) has been studied widely. The mixture of carbon monoxide (CO) gas and hydrogen (H$_2$) gas is called as synthesis gas (syngas), and it can be applied to the synthesis of various chemicals. The production of syngas is conducted at high temperature above 873K and water (H$_2$O) is produced as a by-product. Oxide ceramic membranes such as silica (SiO$_2$) and alumina (Al$_2$O$_3$) which show high thermal resistance have been used as the gas separator, especially as the H$_2$ separator, in the syngas production[1]. In order to improve the efficiency of gas production and separation, many catalyst-loaded membranes have been developed, in which the group VIII metals (Rh, Ru, Ni, Pt, Pd, Ir, Co, Fe) are loaded on the membranes by impregnation, coprecipitation, and so on[1]. However, it has been pointed out that the gas permeability of SiO$_2$ membrane deteriorates by exposure to steam above 823K[2].

Silicon carbide (SiC) ceramics which shows excellent resistance for high temperature, corrosion, and steam is expected to be a candidate alternating material of oxide ceramics. We have developed a SiC film for H$_2$ gas separation which shows H$_2$/N$_2$ perm-selectivity of over 250 at 523K from Si-based polymer utilizing radiation curing[3]. Si-based polymers, typically polycarbosilane (PCS), have a good point of high formability and ability to form a mixture with various materials. We recently have examined development of a catalyst-loaded SiC material from a mixture composed of Si-based polymer and transition metal compounds as a precursor in order to fabricate a SiC film with improved gas perm-selectivity. This approach proposes the synthesis of catalyst-loaded SiC material from the precursor by a lump, which simplifies the synthesis process comparing with conventional methods of

7

impregnation and coprecipitation. In this work, palladium(II) acetate ($Pd(OAc)_2$) was adopted as the transition metal compound, because palladium is a well known catalyst for automotive emissions control, organic synthesis, and so on[4-5]. This paper describes the synthesis and characterization of a novel precursor composed of PCS and $Pd(OAc)_2$ and the catalytic performance of Pd-loaded SiC material obtained by the pyrolysis of novel precursor.

EXPERIMENTAL PROCEDURE

Polycarbosilane (PCS; NIPUSI Type A, purchased from Nippon Carbon Co. Ltd.) is a solid polymer at room temperature and has a number average molecular weight of 2000. Palladium(II) acetate ($Pd(OAc)_2$) is orange-brown powder with molar mass of 224.5g/mol. The chemical structures of PCS and $Pd(OAc)_2$ are represented in Fig. 1. Each PCS and $Pd(OAc)_2$ were dissolved in tetrahydrofuran (THF) of 50ml with concentration of about 2wt%. The $Pd(OAc)_2$ solution was added by dropping to the PCS solution with stirring at ambient temperature. The temperature of solution was monitored during the dropping. After the mixing, THF was evaporated. And then, a precursor composed of PCS and $Pd(OAc)_2$ (PCS-$Pd(OAc)_2$) of black powder was obtained. The ratio of PCS/$Pd(OAc)_2$ was 1/0.8 in weight, where Si-H bond in PCS and carboxylate group (-$OCOCH_3$) in $Pd(OAc)_2$ are equivalent in number.

FT-IR measurement and TGA in He gas atmosphere were conducted for the obtained PCS-$Pd(OAc)_2$. PCS and $Pd(OAc)_2$ were also measured by FT-IR and TGA as references.

Pd-loaded SiC powder was obtained by the pyrolysis of PCS-$Pd(OAc)_2$ at 1373K for 1h in Ar gas atmosphere. The catalytic performance of Pd-loaded SiC powder was evaluated by evolution of $CO_2$ gas due to oxidation of CO gas. The Pd-loaded SiC powder was sealed in the glass tube with 0.2%CO-2%$O_2$-97.8%$N_2$ gas and heat-treated at 473K for 1h. After the heat treatment, the amount of evolved $CO_2$ gas was measured by a gas chromatograph.

Powder XRD measurement of the PCS-$Pd(OAc)_2$ pyrolyzed at 1373K in Ar was conducted in order to investigate the products.

RESULTS AND DISCUSSION

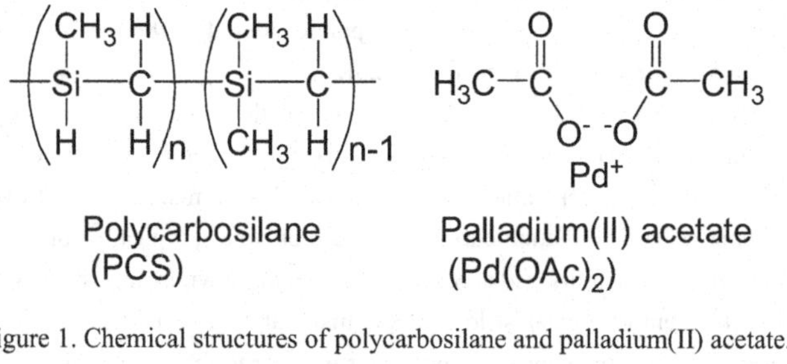

Polycarbosilane
(PCS)

Palladium(II) acetate
($Pd(OAc)_2$)

Figure 1. Chemical structures of polycarbosilane and palladium(II) acetate.

Figure 2 shows the change of temperature of solution from the beginning to the end of the dropping of Pd(OAc)₂ solution. The temperature increased just after the dropping, and reached to +2.3K at 800sec. This suggests that somewhat exothermic reaction occurred between PCS and Pd(OAc)₂ during the mixing.

Figure 3 shows FT-IR spectrum of obtained PCS-Pd(OAc)₂. In the case of PCS, the absorbing

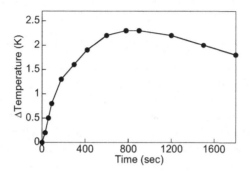

Figure 2. The change of temperature of solution during dropping of Pd(OAc)₂ solution.

peaks assigned to C-H stretching (2950, 2900cm⁻¹), Si-H stretching (2100cm⁻¹), C-H deformation (1460, 1410cm⁻¹), Si-CH₃ stretching (1250cm⁻¹), CH₂ deformation in Si-CH₂-CH₂-Si (1130cm⁻¹), CH₂ deformation in Si-CH₂-Si (1020cm⁻¹), and CH₂ bending of Si-CH₂-Si (800cm⁻¹) were found. In the case of Pd(OAc)₂, the absorbing peaks assigned to C=O asymmetric stretching (around 1560cm⁻¹) and C=O symmetric stretching (1415cm⁻¹) in the ionized carboxylate groups (-OCOCH₃) were found[6]. In the case of PCS-Pd(OAc)₂, it was found that the peak of Si-H (2100cm⁻¹) decreased and a peak assigned to C=O stretching (1716cm⁻¹) in carboxyl group appeared. The appearance of C=O bonds in the

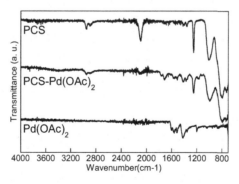

Figure 3. FT-IR spectrum of PCS-Pd(OAc)₂.

PCS-Pd(OAc)$_2$ indicates that PCS reacts with the carboxylate groups in Pd(OAc)$_2$. Furthermore, a peak broadening due to the formation of Si-O-Si and/or Si-O-C bonds was found at 1000-1100cm$^{-1}$. The decrease of Si-H bond and the appearance of Si-O-Si and/or Si-O-C bonds indicate the oxidative crosslinking of molecules in PCS[7]. While it is unclear whether the oxygen source in the oxidative crosslinking is carboxylate groups or dissolved oxygen in THF or others, we suppose that the crosslinking occurs due to a catalytic action of Pd. These reactions between PCS and Pd(OAc)$_2$ are considered to be a reason for the exothermic reaction as shown in Fig. 2.

Figure 4 shows TG curve of obtained PCS-Pd(OAc)$_2$ up to 1473K in He gas atmosphere. In the case of PCS, large mass loss was found in the temperature range of 600-1000K and the mass residue at 1473K was 56%. Such large mass loss is caused by evolution of gaseous products such as hydrogen, methane, and low molecular weight components[8]. In the case of Pd(OAc)$_2$, the mass decreased drastically around 520K and the mass residue at 1473K was 42%. On the other hand, in the case of PCS-Pd(OAc)$_2$, two stage of mass loss were found in the temperature range of 340-480K and 480-1000K. The mass loss in the lower temperature range of 340-480K may be attributed by decomposition of carboxylate groups (-OCOCH$_3$). The mass loss in the temperature range of 480-1000K was much smaller than that of PCS. This is considered to be caused by suppress of evolution of gaseous products such as hydrogen, methane, and low molecular weight components from PCS due to the oxidative crosslinking in PCS. The mass residue of PCS-Pd(OAc)$_2$ at 1473K was 70%.

According to the results as mentioned above, it was found that PCS and Pd(OAc)$_2$ react with

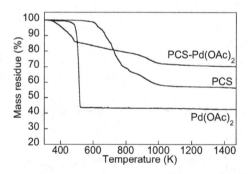

Figure 4. TG curve of PCS-Pd(OAc)$_2$ in He gas atmosphere.

each other when their solution are mixed and that the mass residue of PCS-Pd(OAc)$_2$ after the pyrolysis increases by 14% compared with that of PCS. Pd(OAc)$_2$ acts as a crosslinking regent of PCS and enhances the ceramic conversion yield.

Table I. Evaluation result of catalytic performance for Pd-loaded SiC material.

| Sample | Evolved $CO_2$ gas (mol/g) |
|---|---|
| Pd-loaded SiC material synthesized from PCS-Pd(OAc)$_2$ | $2.4 \times 10^{-6}$ |
| SiC material synthesized from PCS | $7.1 \times 10^{-8}$ |
| Pd material Synthesized from Pd(OAc)$_2$ | $8.5 \times 10^{-5}$ |

The evaluated result of catalytic performance for the obtained Pd-loaded SiC powder is shown in Table I. While $CO_2$ gas hardly evolved from the SiC powder obtained from PCS, $CO_2$ gas evolved with the amount of $2.4 \times 10^{-6}$ mol/g from the Pd-loaded SiC powder obtained from PCS-Pd(OAc)$_2$. On the other hand, in case of Pd material obtained by pyrolysis of Pd(OAc)$_2$ at 573K in Ar, $CO_2$ gas evolved with the amount of $8.5 \times 10^{-5}$ mol/g. According to the result of TGA, the weight ratio of Pd in the Pd-loaded SiC powder becomes 26wt% ideally, so that the amount of $CO_2$ gas evolved from the Pd-loaded SiC powder is estimated to be $2.1 \times 10^{-5}$ mol/g. However, the amount of evolved $CO_2$ gas was really 11% of the estimation. This is considered to be caused by the difference of products after the pyrolysis. Figure 5 shows XRD pattern of pyrolyzed PCS-Pd(OAc)$_2$. Broad peaks assigned to $\beta$-SiC were found in PCS and very sharp peaks assigned to metallic Pd and Pd hydride were found in Pd(OAc)$_2$. On the other hand, in the case of PCS-Pd(OAc)$_2$, peaks assigned to $\beta$-SiC and Pd$_2$Si were found. The production of Pd$_2$Si explain the low catalytic performance of Pd-loaded SiC powder obtained from PCS-Pd(OAc)$_2$, because Pd$_2$Si does not show catalytic performance.

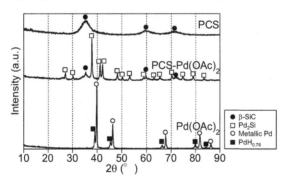

Figure 5. XRD pattern of PCS-Pd(OAc)$_2$ pyrolyzed at 1373K in Ar.

CONCLUSION

In this work, a novel precursor composed of polycarbosilane (PCS) and palladium(II) acetate $(Pd(OAc)_2)$ was synthesized and characterized in order to develop a catalyst-loaded SiC material.

It was found that $Pd(OAc)_2$ acts as a crosslinking regent of PCS leading to enhancement of the ceramic conversion yield and that Pd-loaded SiC material obtained by the pyrolysis of the novel precursor shows a catalytic performance. However, the catalytic performance of obtained Pd-loaded SiC material was low because of the production of $Pd_2Si$. It is a future problem to suppress the production of $Pd_2Si$ in order to improve the catalytic performance.

ACKNOWLEDGMENT

This work was supported by a Grant-in-Aid for Scientific Research from the Japan Society for the Promotion of Science.

REFERENCES

[1] S. Wang, G. Q. Lu, and G. J. Millar, Carbon dioxide reforming of methane to produce synthesis gas over metal-supported catalysts: State of the art, *Energy & Fuels*, **10**, 896-904 (1996).

[2] S. Kim, and G. R. Gavalas, Preparation of H2 permselective silica membranes by alternating reactant vapor deposition, *Ind. Eng. Chem. Res.*, **34**, 168-176 (1995).

[3] R. A. Wach, M. Sugimoto, A. Idesaki and M. Yoshikawa, Molecular sieve SiC-based membrane produced by radiation curing of preceramic polymers for hydrogen separation, *Mater. Sci. Eng. B*, **140**(1-2), 81-89 (2007).

[4] Y. Nishihata, J. Mizuki, T. Akao, H. Tanaka, M. Uenishi, M. Kimura, T. Okamoto, N. Hamada, Self-regeneration of a Pd-perovskite catalyst for automotive emissions control, *Nature*, **418**, 164-165 (2002).

[5] "Palladium in organic synthesis," ed. by J. Tsuji (Springer), (2005).

[6] Q. Fang, G. He, W. P. Cai, J. Y. Zhang, I. W. Boyd, Palladium nanoparticles on silicon by photo-reduction using 172 nm excimer UV lamps, *App. Surf. Sci.*, **226**, 7-11 (2004).

[7] H. Q. Ly, R. Taylor, R. J. Day, F. Heatley, Conversion of polycarbosilane (PCS) to SiC-based ceramic Part1. Characterisation of PCS and curing products, *J. Mater. Sci.*, **36**, 4037-4043 (2001).

[8] Y. Hasegawa, K. Okamura, Synthesis of continuous silicon carbide fibre: Part 3 Pyrolysis process of polycarbosilane and structure of the products, *J. Mater. Sci.*, **18**, 3633-3648 (1983).

# SOLID-STATE NMR STUDIES ON PRECURSOR-DERIVED SI-B-C-N CERAMICS

Otgontuul Tsetsgee[a] and Klaus Müller[a,b,*]
[a]Institut für Physikalische Chemie, Universität Stuttgart, Pfaffenwaldring 55, D-70569
Stuttgart, Germany
[b]Dipartimento di Ingegneria dei Materiali e Tecnologie Industriali, Università degli Studi di Trento, via
Mesiano 77, I-38123, Trento, Italy and
INSTM, UdR Trento, Italy

ABSTRACT
    Precursor-derived Si-B-C-N ceramics are known for their exceptional good thermal stability up
to temperatures of about 2000 °C. It is anticipated that this property is strongly related to the unique
structural composition, comprising nanocrystalline SiC and $Si_3N_4$ domains along with a turbostratic
$BNC_x$ phase. In this contribution, solid-state NMR studies are performed in order to get a deeper
insight into the structural composition of these ceramic materials. Particular emphasis is given to the
$BNC_x$ phase and its alteration during the thermolytic preparation process. In this context, suitable
NMR experiments are presented, from which internuclear boron-nitrogen and boron-boron distances
can be derived. Generally, these internuclear distances are found to be longer than those in pure $h$-BN.
These longer distances are in agreement with other experimental NMR data which also point to
materials with some local disorder, which holds even for the highest annealing temperature. The longer
internuclear distances within the $BNC_x$ component are related to some stress caused by the $sp^2$-carbon
layers within the $BNC_x$ phase. Hence, it is claimed that at 1050 °C and above, distorted BN layers and
$sp^2$-carbon sheets form the $BNC_x$ phase. Whether they occur as intercalated, but separate layers or
whether they exist as interdigitated nanodomains of h-BN and $sp^2$-carbon cannot be conclusively
answered.

INTRODUCTION
    Non-oxide ceramics containing the elements Si, C, N and B exhibit very good chemical and
mechanical stability at elevated temperatures, and are therefore suitable candidates for high
temperature applications.[1-3] Materials, such as $Si_3N_4$, SiC, BN and $Si_3N_4$/SiC composites, are typically
prepared by powder technology,[4,5] where densification of the powders is achieved by sintering at high
temperatures[6], using sintering additives,[7,8] to overcome the low self-diffusion coefficients in these
materials.[9] An alternative approach that recently has attracted wide attention is the precursor route. It
relies on suitable precursor polymers and offers a number of advantages, as compared to the classical
powder technology. In this way, novel ceramic materials with high temperature stability and good
oxidation resistance can be obtained without using additional sintering additives.[6,10-16] For instance, Si-
C-N ceramics were derived via this precursor route, and were found to exhibit exceptional material
properties, as expressed by a high creep and corrosion resistance, high tensile strength and hardness.
However, independent of the specific precursors, their thermal stability is limited to about 1500 °C,
due to the reaction of $Si_3N_4$ with free carbon to form $N_2$.[17-19]
    More recently, it has been demonstrated that the addition of boron to Si-C-N materials increases
substantially the high-temperature stability and shifts the temperature of crystallization towards higher
temperature. Some Si-B-C-N materials remain amorphous up to 1700 °C, and exhibit only minor
decomposition below 2000 °C.[20-25] During recent years numerous Si-B-C-N ceramic systems of
different compositions were synthesized and structurally characterized by techniques, including solid-
state NMR,[26-34] FTIR,[35,36] X-ray diffraction [35,37-39] , TEM[40,41] as well as XPS.[26,42] These studies

primarily addressed the basic precursor structure, the structural evolution and the mechanisms during the thermolytic transformation to the final ceramic materials.

Despite the enormous synthetic[43-62] and characterization efforts, the particular role of boron in these materials is still not completely understood. It has been speculated that the presence of boron might delay the diffusion of other atoms, and thus the nucleation of SiC and $Si_3N_4$ nanocrystallites, by reducing the carbon and nitrogen activities via formation of a $BNC_x$ matrix.[40,63] Hence, it seems that the $BNC_x$ phase plays an important role for the unusual high temperature stability of these materials. However, a satisfactory and complete picture about the structural composition of the $BNC_x$ phase is still missing. There is still lack of evidence how these elements are structurally organized, i.e., in a honeycomb network or as mixtures of separate graphene and h-BN layers. For further developments of ceramics with better material properties it is therefore essential to clarify the structural role of boron in such Si-B-C-N ceramics.

In the present contribution, solid-state NMR techniques are used which primarily address this aforementioned issue. For this purpose, spectral analysis as well as suitable single and double resonance measurements - $^{11}B$ spin echo, $^{11}B\{^{15}N\}$ REDOR (rotational echo double resonance), $^{11}B\{^{14}N\}$ REAPDOR (rotational echo adiabatic passage double resonance) experiments - are performed, the latter of which exploit the dipolar couplings between like and unlike spins to extract boron-boron and boron-nitrogen distances. It is worthwhile to note that REDOR studies using other combinations of nuclei were reported earlier for another Si-B-C-N ceramic system.[64-68] However, in that work only a single $^{13}C$, $^{29}Si$ and $^{15}N$ enriched sample, heat-treated at one particular temperature, was available.[34] Moreover, $^{11}B$ spin echo and $^{11}B\{^{15}N\}$ REDOR echo experiments on two series of Si-B-C-N ceramic systems were also published recently from our group.[34]

In the present work, the structural features of the $BNC_x$ phase are further examined by consideration of additional Si-B-C-N systems. In this context, several $^{15}N$ enriched and non-enriched samples were available, pyrolysed at different temperatures to mimic the structural evolution during the various stages of the ceramic preparation route. In particular, for the first time $^{11}B\{^{14}N\}$ REAPDOR experiments for the derivation of boron-nitrogen distances are presented, which are particularly attractive, since they do not require expensive $^{15}N$ enrichment as for the aforementioned REDOR experiments. From reference measurements on h-BN (whose structure is known[69]), it is shown that the REAPDOR experiment can be also successfully applied for the extraction of reliable distance information is such ceramic systems.

EXPERIMENTAL

Samples

The precursor derived Si-B-C-N ceramics were prepared by thermolysis of the boron-modified polymethylvinylsilazane **1**, polyhydridovinylsilazane **2** and polyhydridomethylaminovinylsilazane **3**, whose chemical structures are given in Figure 1. Details about the precursor synthesis can be found elsewhere.[30-34,37-39,54-56] The $^{15}N$ enriched precursors were synthesized by using $^{15}N$-enriched ammonia (99 at. % $^{15}N$; Campro Scientific, Berlin).

Samples were prepared by pyrolysis of 1 to 2 g of the polymeric precursor in a quartz or aluminium oxide tube under a steady flow (50 ml/min) of purified argon in a programmable tube furnace (Gero HTRV 40-250). Starting at ambient temperature, the following heating program was used: (i) an initial 1 K/min ramp to the desired thermolysis temperature, (ii) a 2 h hold at the thermolysis temperature (samples heat treated at 1400 °C), and (iii) sample cooling with a rate of 2 K/min, during which the sample was allowed to cool to room temperature. Annealing at higher temperatures (1600 – 2000 °C) was done on samples thermolyzed at 1400 °C/2h in argon atmosphere. The heating and cooling rates were 20 K/min to 1400 °C, 2 °C to the desired annealing temperature (dwell time: 5h) and 20 K/min to room temperature.

NMR measurements.

$^{13}$C single pulse MAS NMR experiments were performed on a Bruker CXP 300 spectrometer operating at a static magnetic field of 7 T using a Bruker MAS 4 mm probe and a sample spinning rate of 10 kHz. The 90° pulse width was 4 μs, and the recycle delay was 15 s. $^{13}$C chemical shift were determined relative to the external standard adamantane (δ = 58.36 ppm). The number of scans was 7200.

All other solid-state NMR experiments were carried out on a Varian InfinityPlus 400 NMR spectrometer operating at a static magnetic field of 9.4 T, using a 4 mm triple resonance probe. The resonance frequencies were 28.0, 40.5 and 128.3 MHz, for $^{14}$N, $^{15}$N and $^{11}$B, respectively.

The $^{29}$Si MAS NMR spectra were recorded under MAS condition (sample rotation frequency: 5 kHz) by direct excitation applying 45° pulses of 2.3 μs in width, and recycle delays of 45 s. Chemical shifts were determined relative to the external standard $Q_8M_8$, the trimethylsilylester of octametric silicate (δ = 11 ppm). $^{15}$N MAS NMR spectra were recorded under MAS conditions (sample rotation frequency: 10 kHz) by direct excitation, applying 90° pulses of 3.5 μs, and recycle delays of 15 s. $^{15}$N chemical shifts were referenced to external glycine (100% $^{15}$N-enriched) with a chemical shift of δ = -345 ppm.

$^{11}$B NMR spectra, $^{11}$B{$^{15}$N} REDOR and $^{11}$B{$^{14}$N} REAPDOR experiments (see Fig. 3) were recorded with sample spinning frequency of 10 kHz, recycle delays between 2 and 64 s depending on the $^{11}$B spin-lattice relaxation time. $^{11}$B 90° and 180° pulse widths were 1 and 1.95 μs, respectively. $^{11}$B chemical shifts were determined relative to the external standard B(OH)$_3$ (δ = 18.3 ppm).

The $^{15}$N dephasing 180° pulse width for the REDOR and the $^{14}$N adiabatic pulse width for REAPDOR experiment were 7 and 33.33 μs, respectively. To reach the $^{14}$N resonance frequency, the low gamma box was used for REAPDOR experiments. $^{11}$B spin echo experiments were done on static samples.


Data analysis

a) $^{11}$B spin echo experiments. The homonuclear second moment, $M_{2(homo)}$, characterizing the homonuclear dipole-dipole coupling among the $^{11}$B spins, can be measured by the spin echo pulse sequence (see Figure 2 (left))[70,71]. Ideally, the spin echo intensity during the evolution time, 2τ, is only attenuated by homonuclear dipolar couplings. For short evolution periods in a multispin system, the normalized echo intensity, $I(2τ)/I_0$, can be described by a Gaussian function

$$\frac{I(2\tau)}{I_0} = \exp\left[ -(2\tau)^2 \frac{M_{2(homo)}}{2} \right] \qquad (1)$$

where $M_{2(homo)}$ denotes the second moment of the homonuclear dipolar couplings. In case of spin-3/2 nuclei, the homonuclear second moment associated with the central spin transition can be approximated by[72,73]

$$M_{2(homo)} = 0.9562 \left( \frac{\mu_0}{4\pi} \right)^2 \gamma^4 \hbar^2 \sum_j \left( \frac{1}{r_{ij}^6} \right) \qquad (2)$$

where $r_{ij}$ are the internuclear distances. Equation (1) is fulfilled under the following approximations: (a) the quadrupolar satellite transitions are shifted far off-resonance, (b) the chemical shift differences of the interacting spin pairs are smaller than the strength of the dipolar couplings, and (c) heteronuclear dipolar couplings are negligible on the time scale of the experiment. These conditions are satisfied if the analysis is restricted to the initial part ($0 < 2\tau < 200$ μs) of the spin echo decay.[74,75]

b) $^{11}$B{$^{15}$N}REDOR experiments. The REDOR experiment reintroduces the heteronuclear dipolar interaction between two types of coupled spins (here $^{11}$B and $^{15}$N nuclei) under MAS. The REDOR experiment (see Fig. 3) consists of two parts. The first experiment, a rotor-synchronized spin echo ($90°$ - $\tau$ - $180°$ - $\tau$) sequence applied on the observed nuclei (here $^{11}$B), serves as reference experiment, for which the averaged value of dipolar frequency, $\omega_D$, over each rotor cycle is zero ($\bar{\omega}_D = 0$). In the second experiment, a sequence of $180°$ pulses is applied to the coupled spins (here $^{15}$N) at every half rotor period which inverts the precession direction of the magnetization for the observed nucleus. Therefore, the averaged heteronuclear dipolar frequency over one rotor period becomes non-zero. The accumulated phase angle, $\Phi$, of the magnetization after $N_c$ rotor cycles is given by[76-78]

$$\Phi = \bar{\omega}_D 2 N_c T_r = \frac{2 N_c T_r D}{\pi} \sqrt{2} \sin 2\beta \sin \alpha \qquad (3)$$

where $T_r$ is the rotor period, and the angles $\alpha$ and $\beta$ are the polar angles of the internuclear vector between two interacting (i.e. dipolar coupled) spins with respect to the rotor spinning axis. For a powder sample, the normalized echo amplitude is obtained by averaging over all crystallite orientations according to

$$\Delta S / S_0 = 1 - S / S_0 = 1 - \frac{1}{2\pi} \int_\alpha \int_\beta \cos(\Phi) \sin \beta \, d\beta \, d\alpha \qquad (4)$$

$S_0$ and $S$ are the signal intensities for the experiment without (= reference experiment) and with additional $\pi$ pulses on the indirect nucleus, respectively. It can be shown that the above expression not only depends on the dipolar coupling strength between the interacting spins, manipulated during the REDOR experiment, but also on the number of coupled nuclei (i.e. coordination number) and the respective bonding angles. The experimental REDOR curves can be analyzed numerically on the basis of expressions (3) and (4).

c) $^{11}$B{$^{14}$N}REAPOR experiments. The REAPDOR experiment is designed to recover the heteronuclear dipolar coupling between spin-1/2 and quadrupolar nuclei.[79-84] Like the REDOR technique, the REAPDOR relies on two experiments, i.e., one reference experiment to measure the full spectrum ($S_0$) and a second experiment to get the spectrum ($S$) reduced in intensity due to dipolar dephasing. In both experiments, $180°$ pulses are applied to the observed nuclei (here $^{11}$B) every half rotor period, except at the midpoint of the experiment. In the first experiment, the dipolar dephasing of the two halves of the dipolar evolution period are identical, but opposite in sign. This yields again the reference signal ($S_0$). In the second experiment, the dipolar dephasing between the two coupled spins (here $^{11}$B and $^{14}$N) is reintroduced by a single radio-frequency pulse applied on the quadrupolar nucleus (here $^{14}$N) in the middle of the dipolar evolution period, resulting in the reduced signal, $S$. Effective irradiation on the $^{14}$N nuclei only occurs if an adiabatic-passage pulse is applied, for which the adiabatic passage condition[85]

$$v_I^2 / v_Q v_R > 1 \qquad (5)$$

must be satisfied. In this expression $v_I$, $v_R$ and $v_Q$ are the nutation frequency, the rotor frequency and quadrupolar coupling frequency, respectively. It can be shown that dipolar recoupling becomes maximum, if the adiabatic passage pulse length is about one third of the rotor period $(T_r/3)$.[86] In the present work, for the $^{11}B\{^{14}N\}$ REAPDOR experiments on the Si-B-C-N ceramics, a radio frequency field amplitude of $v_1 = 70$ kHz and a rotor speed of $v_r = 10$ kHz were applied. These parameters fulfil the adiabacity condition, taking into account the $^{14}N$ quadrupolar frequency of 210 kHz, as reported for h-BN.

Numerical simulations for the $^{11}B\{^{15}N\}$ REDOR experiments were done with laboratory written MATLAB routines[34,87] and the SIMPSON program package.[88] From reference calculations, it could be shown that both programs provide the same results. The final REDOR curve analysis was therefore done with the MATLAB program which only accounts for heteronuclear $^{11}B$-$^{15}N$ couplings and different orientations of the dipolar interaction tensors, as described in Ref. 34.

In case of the REAPDOR experiments, the theoretical curves were calculated with the SIMPSON program package, by considering a 4-spin system, $BN_3$, the quadrupolar interaction of the $^{14}N$ nuclei, the $^{11}B$-$^{14}N$ dipolar coupling, the orientations of the respective interaction tensors and the strength of the RF field, $v_1$.

RESULTS AND DISCUSSION

In the present work, Si-B-C-N ceramics derived from hydroborated polysilazanes, $\{B[C_2H_4Si(R)X]_3\}$ with $R = CH_3$ and $X = NH$ (**1**), $R = H$ and $X = NH$ (**2**), $R = H$ and $X = NCH_3$ (**3**), were examined by solid-state NMR spectroscopy. It should be noted that the samples of the precursor systems **1** and **2** were 99 atom % $^{15}N$ enriched.

The $^{13}C$, $^{29}Si$ and $^{15}N$ MAS NMR studies showed that the structures of the derived ceramics depend on the elemental composition, and on the groups X and R which are directly involved in the cross-linking and molecular rearrangement processes during the thermolytic precursor-to-ceramic conversion.

To begin with, some representative NMR spectra from these ceramic samples, exposed to a pyrolysis temperature of 1400 °C, will be presented and discussed. Figure 4 compares the $^{29}Si$ MAS NMR spectra of the pyrolysis intermediates **1**, **2** and **3** at 1400 °C. All spectra exhibit broad resonances, as expected for such systems with an amorphous Si-C-N network. They reflect a distribution of bond lengths and angles as well as chemical variations in the coordination spheres. Deconvolution of the experimental spectra were done, accounting for different $SiC_xN_{4-x}$ (x = 0, 1, 2, 3 and 4) units and using the chemical shift values from literature (see Table 1). This analysis clearly shows that the pyrolysis intermediate **3** (with $X = NCH_3$) exhibits the highest amount of nitrogen-rich silicon sites ($SiCN_3$ and $SiN_4$ units), while for intermediate **1** (with $R = CH_3$) the carbon-rich silicon sites ($SiC_3N$ and $SiC_4$ units) dominate.

The respective $^{13}C$ MAS NMR spectra of these samples are given in Figure 5. The resonances in the range from 0 to 50 ppm stem from $sp^3$-carbon, i.e., $CSi_4$ units in the amorphous Si-C-N network. The resonances in the range from 100 to 150 ppm are attributed to $sp^2$-carbons. The signal at 120 ppm is due to an amorphous carbon phase with a graphite-like structure. The nature of the $sp^2$-carbons at 138 ppm, clearly visible for precursor system **3**, is not yet fully understood. From other solid-state NMR studies, this latter low-field shifted $sp^2$-carbon resonance at around 140 ppm has been attributed to carbons bonded to heteronuclei, such as nitrogen or boron. For the present systems the bonding to boron can be excluded, as in the $^{11}B$ NMR spectra (see below) such a signal component cannot be observed. It therefore seems that the $^{13}C$ resonance at 138 ppm reflects a $sp^2$-carbon-nitrogen bond, either in the $BNC_x$ or SiCN network. However, further investigations are necessary for final clarification.

A further comparison of the present spectra shows that the relative amount of $sp^2$ carbon is highest for precursor system **3**, which also contains the highest content of $SiN_4$ units (see above). At the same time, the strongest $sp^2$-carbon peak is seen for precursor system **1**, where – as discussed above – the $^{29}Si$ NMR spectra is dominated by the carbon-rich silicon sites. Accordingly, the amount of the nitrogen-rich silicon sites is correlated with the $sp^2$-carbon content.

The $^{15}N$ NMR spectra of the two $^{15}N$ enriched precursor systems **1** and **2**, shown in Figure 6, confirm the above results. The $^{15}N$ NMR spectrum of precursor **1** was found to consist of three structural units, $NHB_2$ (25 %), $NB_3$ (37 %) and $NSi_3$ (38 %), while precursor **2** exhibits only $NB_3$ and $NSi_3$ units with relative intensities of 75 and 25 %, respectively (see Table 2). It proves that the cross-linking of the precursor **2**, with a highly reactive Si-H bond in the original precursor, is more efficient, resulting in an increase of the $BN_3$ units as compared to precursor system **1**. It can be also seen that the relative amount of the $NSi_3$ units in precursor system **1** is higher than in precursor system **2**, which is in line with the $^{29}Si$ NMR data, as discussed above.

In order to learn more about the incorporation of boron in the present Si-B-C-N ceramics, we have undergone a comprehensive $^{11}B$ NMR study which will be briefly outlined next. The $^{11}B$ MAS NMR spectra of precursor system **1**, after thermal annealing at temperatures between 200 and 1400 °C, are presented in Figure 7. For comparison, the $^{11}B$ NMR spectrum of h-BN is also given. The $^{11}B$ NMR line shape of $h$-BN shows a characteristic second-order broadening due to the strong quadrupolar interaction of the $^{11}B$ nucleus in this compound (quadrupolar coupling constant $C_{qcc}$ = 2.9 MHz, an asymmetry parameter $\eta$ = 0, isotropic chemical shift $\delta_{iso}$ = 30 ppm). As can be seen, none of the $^{11}B$ NMR spectra of precursor system **1** exhibits the same characteristic $^{11}B$ NMR lineshape as found for $h$-BN. This observation clearly demonstrates that the $^{11}B$ nuclei occur in different environments (i.e. changes in the coordination, bond lengths and angles), as expected in amorphous systems. In general, the $^{11}B$ NMR spectra of precursor system **1** are characterized by tri-coordinated boron atoms, reflecting $BC_xN_{3-x}$ units. The $^{11}B$ isotropic chemical shift values, $\delta_{iso}$, for the possible $BC_xN_{3-x}$ (x = 0, 1, 2, 3) units, obtained from liquid-state NMR studies on suitable reference compounds are summarized in Table 3.

Deconvolution of these experimental $^{11}B$ NMR spectra is impossible. As mentioned above, $^{11}B$ nuclei in h-BN possess an axially symmetric quadrupole coupling tensor ($\eta$ = 0). Therefore, the $^{11}B$ MAS NMR line shape of the central transition exhibits two singularities. The position of the low-field singularity is taken as $\delta_{sing}$. The large quadrupolar coupling of $^{11}B$ in h-BN causes also a second-order high-field shift of the centre of gravity of the respective $^{11}B$ NMR spectrum. For h-BN and our experimental conditions, the centre of gravity is shifted from the true isotropic shift, $\delta_{iso}$, by -13 ppm, while $\delta_{sing}$ occurs at about 20 ppm, and is shifted from $\delta_{iso}$ by -10 ppm.

For this reason, the various spectral components are only discussed on a qualitative basis, by considering the position of the low-field singularity. Accordingly, the low-field shoulders visible in the $^{11}B$ NMR spectra of the 1400, 1050, 400 and 200 °C samples are taken as $\delta_{sing}$ from the $BN_3$, $BCN_2$, $BC_2N$ and $BC_3$ units, respectively. These signal positions are marked by dotted lines in the experimental $^{11}B$ NMR spectra, and are given in Table 3.

The $^{11}B$ NMR spectrum of the 200 °C sample shows resonances due to the carbon enriched $BC_3$ (68 ppm) and $BC_2N$ (45 ppm) units. This indicates that the decomposition of the B-C bonds at this temperature is not complete. In the 400 °C sample, the $BC_3$ units are completely absent, while the intensity of the $BC_2N$ units is reduced and additional high-field resonances show up. They are attributed to $BCN_2$ (at 30 ppm) and $BN_3$ (at 22 ppm) units. After annealing above 600 °C, the $^{11}B$ NMR spectra are dominated by a signal which is characteristic of $BN_3$ units. Obviously, the transformation of the $BC_3$ units into the $BN_3$ units occurs mainly between 200 and 600 °C. The signal from the $BCN_2$ units most likely exists up to 1050 °C. After annealing at 1400 °C the boron atoms are only trigonally coordinated by nitrogen atoms. It should be emphasized that precursor systems **2** and **3** display quite similar $^{11}$B NMR spectra (not shown) as a function of the annealing temperature.

In order to learn more about the structural features in the vicinity of the boron nuclei, double resonance as well as $^{11}$B spin echo experiments were performed, from which internuclear distances can be derived. Hence, $^{11}$B{$^{15}$N} REDOR experiments were performed on the $^{15}$N enriched precursor systems **1** and **2**, covering the temperature range up to 1400 °C. Likewise, $^{11}$B{$^{14}$N} REAPDOR experiments were performed on precursor system **3** and, as reference, on commercial h-BN.

Representative spectra from these REDOR and REAPDOR experiments are shown in Figure 8. It should be noted that the intensities of the dephased spectra (S) in the REDOR and REAPDOR experiments decrease due to dipolar interactions between the detected $^{11}$B nucleus and the surrounding $^{15}$N and $^{14}$N nuclei, respectively.

$^{11}$B{$^{15}$N} REDOR dephasing curves for precursor system **1** are depicted in Figure 9. For the sample annealed at 200 °C no dephasing curve is given, since – as mentioned above – at this temperature boron is primarily bonded to carbon. As mentioned earlier, it was not possible to deconvolute the $^{11}$B NMR spectra into the various components. For this reason, the given experimental dephasing curves were obtained by taking the signal intensity from the dephasing spectra ($\Delta S = S_0 - S$, see equation (4)) at that position, which corresponds to the low-field singularity of the BN$_3$ component. In addition, the derived $^{11}$B{$^{14}$N} REAPDOR dephasing curves for h-BN and precursor system **3** are also shown in Figure 9.

The given theoretical simulations in this figure were made with the assumption that all boron atoms are surrounded by three nitrogen atoms with a N-B-N angle of 120°, as in h-BN. The only variable was the bond length, and a scaling factor which is a measure for the relative amount of BN$_3$ units in the samples. In general, a good match between experimental and theoretical curves could be achieved, which supports the underlying model assumptions.

The obtained parameters from the analysis of these REDOR and REAPDOR experiments are summarized in Table 4. It should be emphasized that the derived boron-nitrogen distance from the $^{11}$B{$^{14}$N} REAPDOR experiment on commercial h-BN is identical to the value of 1.44 Å, as reported from X-ray diffraction studies.[69] In general, both the REDOR and the REAPDOR experiments show a decrease of the boron-nitrogen distances with increasing annealing temperature. From the REDOR experiments, for precursor system **1** distances between 1.75 and 1.60 Å were derived, while precursor system **2** shows values between 1.68 and 1.55 Å in the temperature range of 400 to 1400 °C. Hence, the boron-nitrogen distances in both precursor systems are longer than in h-BN. The boron-nitrogen distances, derived from the REAPDOR experiments on precursor system **3** at 1050 and 1400 °C are somewhat shorter than those of precursor systems **1** and **2** at the same temperatures.

Figure 10 compares the REDOR results for the intermediates **1** and **2** as a function of the annealing temperature. It is seen that at lower temperatures for precursor system **2** the final plateau value of the REDOR curves, i.e. scaling factors, lies below 1. These lower value reflects the presence of the BCN$_2$ units, since the chosen resonance position, from which the REDOR dephasing curve was calculated (see above), also contains signal intensities from the BCN$_2$ units. However, at higher pyrolysis temperatures the maximum value of dipolar dephasing is reached for this precursor system.

Figure 10 also shows that for precursor system **1** the scaling factor does not reach the theoretical maximum for complete dephasing, as expected for a sample with only BN$_3$ units, even at the highest pyrolysis temperatures. This finding indicates that for this precursor system BCN$_2$ units might exist even at the highest pyrolysis temperature. At the same time, precursor system **1** also shows longer boron-nitrogen distances than precursor system **2**, pointing to a better local order in the BN sheets of the latter precursor material. This result is also confirmed by the $^{15}$N NMR results, discussed above, which show for precursor system **1** another structural component apart from the expected NSi$_3$ and NB$_3$ units. Nevertheless, it should be mentioned that in the $^{11}$B NMR spectra of precursor system **1** there is no unequivocal proof for the existence of, for instance, a $BCN_2$ unit. However, one major problem is the general broadening of the $^{11}B$ NMR lineshapes for the present ceramic systems, as discussed above.

$^{11}B$ spin echo experiments were performed for the reference compound h-BN and for precursors **1** to **3**. Figure 11 pesents the experimental and fitted $^{11}B$ spin echo decay curves for h-BN and for the pyrolysis intermediate **1** at 1050 °C. As outlined in the experimental part, the theoretical fit can be used to calculate, via the homonuclear second moment $M_{2(homo)}$, the average boron-boron distance in the BN domains. The distances given in Table 5 were derived by assuming that each boron has 6 next neighbouring borons at the same distance. Here, for the ceramic systems only the values at 1050 °C and 1400 °C are reported. At lower pyrolysis temperatures, the analysis of the spin echo curves does not make sense, as the spin echo experiments are done for static samples and at lower pyrolysis temperatures there is too much overlap from spectral components other than $BN_3$ units.

Inspection of Table 5 reveals that the boron-boron distance of 2.51 Å found for commercial h-BN is identical with the value from X-ray diffraction.[69] For all ceramic samples larger boron-boron distances are calculated which is in agreement with the above results for the boron-nitrogen distances from the REDOR and REAPDOR studies. That is, both the spin echo and the REDOR/REAPDOR experiments clearly prove that the local order in the $BNC_x$ domains is distorted. Although both boron and nitrogen are trigonally coordinated by nitrogen and boron atoms, respectively, the internuclear distances are longer than in pure h-BN. For the ceramic systems **1** and **3** it is further seen that the boron-boron distances decrease from 1050 to 1400 °C which indicates that the local structure approaches the h-BN structure. However, in particular the boron-boron distances are still longer than for the h-BN reference which indicates a lack of intermediate and long-range order in these precursor derived ceramics.

CONCLUSIONS

Precursor-derived Si-B-C-N ceramic systems were studied by solid-state NMR spectroscopy. The investigations presented here primarily address the structural composition of the $BNC_x$ phase which plays a key role in the discussion of the high temperature stability of precursor-derived Si-B-C-N ceramics. Single and double resonance solid-state NMR experiments were employed, from which internuclear distances could be derived. $^{11}B$ spin-echo experiments were employed to obtain boron-boron distances. Likewise, $^{11}B\{^{15}N\}$ REDOR on $^{15}N$ enriched samples and – for the first time - $^{11}B\{^{14}N\}$ REAPDOR experiments were used to get boron-nitrogen distances. As proven by the reference measurements on h-BN, these methods in general provide reliable data. The studies on the ceramics demonstrate that the $BNC_x$ phase is not conform with the ideal structure of h-BN. Obviously, both the boron and nitrogen atoms are trigonally coordinated by nitrogen and boron atoms, respectively, as in h-BN after sample pyrolysis at 1050 °C and above. The ideal BN layer structure, however, is distorted, as reflected by the longer internuclear distances. At the same time, $sp^2$-carbon from graphene sheets can be detected. At present it is open whether these two components, h-BN and $sp^2$-carbon, exist as intercalated, but separate layers or whether they have to be described by interdigitated nanodomains. For instance, a $^{13}C$ resonance observed for one of the presently studied ceramic materials indicates the presence of $sp^2$-carbon-nitrogen bonds, being in line with the latter model.

On the other hand, the model of intercalated boron nitride and (amorphous) $sp^2$-carbon layers might explain the longer interatomic distances, since the graphite-like carbon layers might create some internal pressure which in turn is responsible for the BN layer distortion. However, further work is required to provide a comprehensive structural model for the $BNC_x$ phase in such ceramic systems.

ACKNOWLEDGEMENTS
The authors would like to thank Dr. M. Weinmann and Prof. F. Aldinger (Max-Planck-Institut für Metallforschung, Stuttgart) for providing the ceramic materials as well as Prof. G. Buntkowsky (TU Darmstadt) for the REDOR simulation program. Financial support for this research project by the Deutsche Forschungsgemeinschaft is gratefully acknowledged. O. T. acknowledges financial support by the Deutsche Forschungsgemeinschaft and the Graduiertenkolleg No. 448 "Modern Methods of Magnetic Resonance in Materials Science".

REFERENCES

[1]    E. M. Lenoe, R. N. Katz, J. J. Burke, Ceramics for high performance applications I{I}{I}. Plenum Press: New York, (1983).
[2]    S. Hampshire, in Proc. Int. Conf, Elsevier Applied Science: London, 1986.
[3]    H. Tanaka, P. Greila, G. Petzowa, Sintering and strength of silicon nitride-silicon carbide composites Int. J. High. Tech. Ceram. 1, (2), 107 (1985).
[4]    A. W. Weimer, Carbide, nitride, and boride materials synthesis and processing. Chapman \& Hall: London, (1997).
[5]    D. Segal, Chemical synthesis of advanced ceramic materials. Cambridge University Press: Cambridge, (1989).
[6]    D. Seyferth, G. H. Wiseman, High-yield synthesis of silicon nitride/silicon carbide ceramic materials by pyrolysis of a novel polyorganosilazane J. Am. Ceram. Soc. 67, (7), C132 (1984).
[7]    R. W. Rice, Ceramic fabrication technology. Marcel Dekker: New York, (2003).
[8]    S. Saito, S. Somiya, in International Symposium on Factors in Densification and Sintering of Oxide and Non-oxide Ceramics, Gakujutsu Bunken Fukyo-kai: Hakone Japan, 1987.
9     G. G. Gnesin, Basic Trends in Research and Development of Non-Oxide Ceramic Materials Refract. Ind. Ceram. 41, 155 (2000).
[10]   R. L. K. Matsumoto, J. M. Schwark, Preceramic polymers incorporating boron Patent, US5206327 (1993).
[11]   R. M. Laine, A. Sellinger, in The chemistry of organic silicon compounds, eds. Rappoport, Z.; Apeloig, Y., John Wiley \& Sons: Chichester, 1998; pp 2245.
[12]   H. P. Baldus, M. Jansen, Moderne Hochleistungskeramiken - amorphe anorganische Netzwerke aus molekulaten Vorlaeufern Angew. Chem. 109, (4), 338 (1997).
[13]   A. Kienzle, J. Bill, F. Aldinger, R. Riedel, Nanosized Si-C-N powder by pyrolysis of highly crosslinked silylcarbodiimide Nanostruct. Mater. 6, (1-4), 349 (1995).
[14]   M. Peuckert, T. Vaahs, M. Brück, Ceramics from organometallic polymers Adv. Mater. 2, 398 (1990).
[15]   J. Bill, F. Aldinger, Precursor-derived covalent ceramics Adv. Mater. 7, (9), 775 (1995).
[16]   D. Galusek, S. Reschke, R. Riedel, W. Dreszler, In-Situ Carbon Content Adjustment in Polysilazane Derived Amorphous SiCN Bulk Ceramics J. Eur. Ceram. Soc. 19, (10), 1911 (1999).
[17]   M. Friess, J. Bill, F. Aldinger, D. V. Szabo, Crystallization of Amorphous Silicon Carbonitride Investigated by Transmission Electron Microscopy (TEM) Mech. Corr. Prop. A, (89/91), 95 (1994).
[18]   J. Bill, J. Seitz, G. Thurn, J. Durr, J. Canel, B. Z. Janos, A. Jalowiecki, D. Sauter, S. Schempp, H. P. Lamparter, J. Mayer, F. Aldinger, Structure analysis and properties of Si-C-N ceramics derived from polysilazanes Phys. Status Solidi A 166, (1), 269 (1998).
[19]   Y. Iwamoto, W. Volger, E. Kroke, R. Riedel, T. Saitou, K. Matsunaga, Topical Issue -- Ultrahigh-Temperature Ceramics - Crystallization Behavior of Amorphous Silicon Carbonitride Ceramics Derived from Organometallic Precursors J. Am. Ceram. Soc. 84, (10), 2170 (2001).

[20]    R. Kumar, R. Mager, F. Phillipp, A. Zimmermann, G. Rixecker, High-temperature deformation behavior of nanocrystalline precursor-derived Si-B-C-N ceramics in controlled atmosphere Int. J. Mater. Res. 97, (5), 626 (2006).

[21]    R. Riedel, A. Kienzle, W. Dressler, L. Ruwisch, J. Bill, F. Aldinger, A silicoboron carbonitride ceramic stable to 2000 DegC Nature 382, (6594), 796 (1996).

[22]    Z.-C. Wang, F. Aldinger, R. Riedel, Novel silicon-boron-carbon-nitrogen materials thermally stable up to 2200 DegC J. Am. Ceram. Soc. 84, (10), 2179 (2001).

[23]    Z.-C. Wang, P. Gerstel, G. Kaiser, J. Bill, F. Aldinger, Synthesis of Ultrahigh-Temperature Si-B-C-N Ceramic from Polymeric Waste Gas J. Am. Ceram. Soc. 88, (10), 2709 (2005).

[24]    M. Takamizawa, Patent, US4505151 (1985).

[25]    M. Takamizawa, T. Kobayashi, A. Hayashida, Y. Takeda, Patent, US4604367 (1986).

[26]    Q. D. Nghiem, J.-K. Jeon, L.-Y. Hong, D.-P. Kim, Polymer derived Si-C-B-N ceramics via hydroboration from borazine derivatives and trivinylcyclotrisilazane J. Organomet. Chem. 688, (1), 27 (2003).

[27]    C. Gerardin, F. Taulelle, D. Bahloul, Pyrolysis chemistry of polysilazane precursors to silicon carbonitride.2. Solid-state NMR of the pyrolytic residues J. Mater. Chem. 7, (1), 117 (1997).

[28]    C. Gervais, F. Babonneau, L. Ruwisch, R. Hauser, R. Riedel, Solid-state NMR investigations of the polymer route to SiBCN ceramics Can. J. Chem. 81, (11), 1359 (2003).

[29]    W. R. Schmidt, D. M. Narsavage-Heald, D. M. Jones, P. S. Marchetti, D. Raker, G. E. Maciel, Poly(borosilazane) precursors to ceramic nanocomposites Chem. Mater. 11, (6), 1455 (1999).

[30]    J. Schuhmacher, M. Weinmann, J. Bill, F. Aldinger and K. Müller, Solid State NMR Studies of the Preparation of Si-C-N Ceramics from Polysilylcarbodiimide Polymers, Chem. Mater. 10, 3913 (1998)

[31]    K. Müller, Solid State NMR investigations for Ceramic Characterization in "Grain Boundary Dynamics of Precursor-Derived Covalent Ceramics", VCH, Weinheim (1999), p. 197

[32]    J. Schuhmacher, F. Berger, M. Weinmann, J. Bill, F. Aldinger and K. Müller, Solid State NMR Studies and FT IR Studies of the Preparation of Si-B-C-N Ceramics from Polysilazane Precursors Appl. Organomet. Chem. 15, 809 (2001)

[33]    F. Berger, M. Weinmann, F. Aldinger, K. Müller, Solid State NMR Studies of the Preparation of Si-Al-C-N Ceramics from Aluminium-modified Polysilazanes and Polysilylcarbodiimides, Chem. Mater. 17, 919 (2004).

[34]    T. Emmler, O. Tsetsgee, G. Buntkowsky, M. Weinmann, F. Aldinger, K. Müller, 11B{15N} REDOR and 11B spin echo studies for structural characterization of Si-B-C-N precursor ceramics, Soft Materials 4, 207 (, 2004)

[35]    S. Bernard, M. Weinmann, P. Gerstel, P. Miele, F. Aldinger, Boron-modified polysilazane as a novel single-source precursor for SiBCN ceramic fibers: synthesis, melt-spinning, curing and ceramic conversion J. Mater. Chem. 15, (2), 289 (2005).

[36]    T. Wideman, E. Cortez, E. E. Remsen, G. A. Zank, P. J. Carroll, L. G. Sneddon, Reactions of Monofunctional Boranes with Hydridopolysilazane: Synthesis, Characterization, and Ceramic Conversion Reactions of New Processible Precursors to SiNCB Ceramic Materials Chem. Mater. 9, (10), 2218 (1997).

[37]    M. Horz, A. Zern, F. Berger, J. Haug, K. Müller, F. Aldinger, M. Weinmann, Novel polysilazanes as precursors for silicon nitride/silicon carbide composites without "free" carbon J. Eur. Ceram. Soc. 25, (2-3), 99 (2005).

[38]    A. Müller, J. Q. Peng, H. J. Seifert, J. Bill, F. Aldinger, Si-B-C-N ceramic precursors derived from dichlorodivinylsilane and chlorotrivinylsilane. 2. Ceramization of polymers and high-temperature behavior of ceramic materials Chem. Mater. 14, (8), 3406 (2002).

[39]    P. Gerstel, A. Müller, J. Bill, F. Aldinger, Synthesis and High-Temperature Behavior of Si/B/C/N Precursor-Derived Ceramics without "Free Carbon" Chem. Mater. 15, (26), 4980 (2003).

[40]   A. Jalowiecki, J. Bill, F. Aldinger, J. Mayer, Interface characterization of nanosized B-doped $Si_3N_4$/SiC ceramics Composites Part A 27, (9), 717 (1996).

[41]   A. Zimmermann, A. Bauer, M. Christ, Y. Cai, F. Aldinger, High-temperature deformation of amorphous Si-C-N and Si-B-C-N ceramics derived from polymers Acta Mater. 50, (5), 1187 (2002).

[42]   R. Franke, S. Bender, I. Arzberger, J. Hormes, M. Jansen, H. Juengermann, J. Loeffelholz, The determination of local structural units in amorphous SiBN3C by means of x-ray photoelectron and x-ray absorption spectroscopy Fresenius. J. Anal. Chem. 354, (7-8), 874 (1996).

[43]   H. Noeth, Decomposition of the Si-N bond by Lewis-acidic boron compounds Z. Naturforsch., 16B, (No. 9), 618 (1961).

[44]   D. Seyferth, H. Plenio, Borasilazane Polymeric Precursors for Boronsilicon Nitride J. Am. Ceram. Soc. 73, 2131 (1990).

[45]   R. Riedel, Appl. Organomet. Chem., 241 (1996).

[46]   J. Bill, A. Kienzle, M. Sasaki, R. Riedel, F. Aldinger, Novel routes for the synthesis of materials in the quaternary system Si-B-C-N and their characterization Adv. Sci. Technol. 3B, (Ceramics: Charting the Future), 1291 (1995).

[47]   H. P. Baldus, G. Passing, D. Sporn, A. Thierauf, Si-B-(N,C): A New Ceramic Material for High-Performance Applications Ceram. Trans. 58, 75 (1995).

[48]   H. P. Baldus, M. Jansen, O. Wagner, New materials in the system Si-(N,C)-B and their characterization Key Eng. Mater. 89-91, (Silicon Nitride 93), 75 (1994).

[49]   H. Juengermann, M. Jansen, Synthesis of an extremely stable ceramic in the system Si/B/C/N using 1-(trichlorosilyl)-1(dichloroboryl)ethane as a single-source precursor Mater. Res. Innovations 2, (4), 200 (1999).

[50]   T. Jaschke, M. Jansen, Improved durability of Si/B/N/C random inorganic networks J. Eur. Ceram. Soc. 25, (2-3), 211 (2005).

[51]   H.-P. Baldus, M. Jansen, Novel high-performance ceramics - amorphous inorganic networks from molecular precursors Angew. Chem., Int. Ed. 36, (4), 328 (1997).

[52]   M.Jansen, Mat. Res.Soc. Symp, 821 (1992).

[53]   T. Jaschke, M. Jansen, A new borazine-type single source precursor for Si/B/N/C ceramics J. Mater. Chem. 16, (27), 2792 (2006).

[54]   S. Bernard, M. Weinmann, D. Cornu, P. Miele, F. Aldinger, Preparation of high-temperature stable Si-B-C-N fibers from tailored single source polyborosilazanes J. Eur. Ceram. Soc. 25, (2-3), 251 (2005).

[55]   M. Weinmann, R. Haug, J. Bill, F. Aldinger, J. Schuhmacher, K. Müller, Boron-containing polysilylcarbodi-imides: a new class of molecular precursors for Si-B-C-N ceramics J. Organomet. Chem. 541, (1-2), 345 (1997).

[56]   A. Müller, P. Gerstel, M. Weinmann, J. Bill, F. Aldinger, Si-B-C-N ceramic precursors derived from dichlorodivinylsilane and chlorotrivinylsilane. 1. Precursor synthesis Chem. Mater. 14, (8), 3398 (2002).

[57]   W. Verbeek, Shaped articles of silicon carbide and silicon nitride Patent, (DE Patent 2218960), DE2218960 (1973).

[58]   L. Ruwisch, in, Darmstadt University of Technology, Germany: 1998; p thesisRuwisch.

[59]   M. Jansen, B. Jaeschke, T. Jaeschke, Amorphous multinary ceramics in the Si-B-N-C system Struct. Bond. 101, (High Performance Non-Oxide Ceramics I), 137 (2002).

[60]   Q. D. Nghiem, D.-P. Kim, Polymerization of Borazine with Tetramethyldivinyldisilazane as a New Class SiCBN Preceramic Polymer J. Ind. Eng. Chem. 12, (6), 905 (2006).

[61]   J. Haberecht, R. Nesper, H. Grutzmacher, A construction kit for Si-B-C-N ceramic materials based on borazine precursors Chem. Mater. 17, (9), 2340 (2005).

[62]   A. Kienzle, A. Obermeyer, R. Riedel, F. Aldinger, A. Simon, Synthesis and structure of the first oligomeric cyclic dimethylsilyl-substituted carbodiimide Chem. Ber. 126, (12), 2569 (1993).

63    A. Jalowiecki, in, Max-Planck-Inst. f\"ur Metallforschung, Stuttgart, Germany: Stuttgart, 1997; p thesisJalowiecki.

64    G. Jeschke, M. Kroschel, M. Jansen, A magnetic resonance study on the structure of amorphous networks in the Si-B-N(-C) system J. Non-Cryst. Solids 260, (3), 216 (1999).

65    L. Van Wüllen, M. Jansen, Random inorganic networks: a novel class of high-performance ceramics J. Mater. Chem. 11, (1), 223 (2001).

66    L. Van Wüllen, U. M\uller, M. Jansen, Understanding Intermediate-Range Order in Amorphous Nitridic Ceramics: A 29Si{11B} REDOR/REAPDOR and 11B{29Si} REDOR Study Chem. Mater. 12, (8), 2347 (2000).

67    Y. H. Sehlleier, A. Verhoeven, M. Jansen, Observation of direct bonds between carbon and nitrogen in SI-B-N-C ceramic after pyrolysis at 1400C Angew. Chem., Int. Ed. 47, (19), 3600 (2008).

68    Y. Sehlleier, A. Verhoeven, M. Jansen, NMR studies of short and intermediate range ordering of amorphous Si-B-N-C-H pre-ceramic at the pyrolysis stage of 600 degreeC J. Mater. Chem. 17, (40), 4316 (2007).

69    R. S. Pease, An x-ray study of boron nitride Acta Cryst. 5, 356 (1952).

70    D. Y. Han, H. Kessemeier, Second-order quadrupolar echo in solids Phys. Rev. Lett. 67, (3), 346 (1991).

71    A. C. Kunwar, G. L. Turner, E. Oldfield, Solid-state spin-echo Fourier transform NMR of potassium-39 and zinc-67 salts at high field J. Magn. Reson. 69, (1), 124 (1986).

72    J. Haase, E. Oldfield, Spin-echo behavior of nonintegral-spin quadrupolar nuclei in inorganic solids J. Magn. Reson., Ser A 101, (1), 30 (1993).

73    B. Gee, H. Eckert, Na-23 Nuclear-Magnetic-Resonance Spin-Echo Decay Spectroscopy of Sodium-Silicate Glasses and Crystalline Model Compounds Solid State Nucl. Magn. Reson. 5, (1), 113 (1995).

74    B. Gee, Homonuclear vanadium-51 dipolar couplings in inorganic solids obtained via Hahn spin echo decay NMR spectroscopy Solid State Nucl. Magn. Reson. 19, (3/4), 73 (2001).

75    L. Van Wüllen, B. Gee, L. Zuechner, M. Bertmer, H. Eckert, Connectivities and Cation Distributions in Oxide Glasses: New Results from Solid State NMR Ber. Bunsen Ges. 100, (9), 1539 (1996).

76    T. Gullion, Introduction to rotational-echo, double-resonance NMR Concepts Magn. Reson. 10, (5), 277 (1998).

77    T. Gullion, Measurement of Heteronuclear Dipolar Interactions by Rotational-Echo, Double-Resonance Nuclear Magnetic Resonance Magn. Reson. Rev. 17, (2), 83 (1997).

78    C. P. Jaroniec, B. A. Tounge, C. M. Rienstra, J. Herzfeld, R. G. Griffin, Recoupling of heteronuclear dipolar interactions with rotational-echo double-resonance at high magic-angle spinning frequencies J. Magn. Reson. 146, (1), 132 (2000).

79    T. Gullion, A. J. Vega, Measuring heteronuclear dipolar couplings for I = 1/2, S > 1/2 spin pairs by REDOR and REAPDOR NMR Prog. Nucl. Magn. Reson. Spectrosc. 47, (3-4), 123 (2005).

81    M. Kalwei, H. Koller, Quantitative comparison of REAPDOR and TRAPDOR experiments by numerical simulations and determination of H-Al distances in zeolites Solid State Nucl. Magn. Reson. 21, (3/4), 145 (2002).

82    E. Hughes, T. Gullion, A. Goldbourt, S. Vega, A. J. Vega, Internuclear distance determination of S=1, I=1/2 spin pairs using REAPDOR NMR J. Magn. Reson. 156, (2), 230 (2002).

83    T. Gullion, Measurement of dipolar interactions between spin-1/2 and quadrupolar nuclei by rotational-echo, adiabatic-passage, double-resonance NMR Chem. Phys. Lett. 246, (3), 325 (1995).

84    C. P. Grey, W. S. Veeman, A. J. Vega, Rotational-echo, nitrogen-14/carbon-13/proton triple resonance, solid-state, nuclear magnetic resonance: a probe of carbon-13-nitrogen-14 internuclear distances J. Chem. Phys. 98, (10), 7711 (1993).

[85]    C. P. Grey, A. J. Vega, Determination of the Quadrupole Coupling Constant of the Invisible Aluminum Spins in Zeolite HY with $^1H/^{27}Al$ TRAPDOR NMR J. Am. Chem. Soc. 117, (31), 8232 (1995).
[86]    R. W. Glaser, A. S. Ulrich, Susceptibility corrections in solid-state NMR experiments with oriented membrane samples. Part I: applications J. Magn. Reson. 164, (1), 104 (2003).
[87]    J. M. Goetz, J. Schaefer, REDOR dephasing by multiple spins in the presence of molecular motion J. Magn. Reson. 127, (2), 147 (1997).
[88]    M. Bak, J. T. Rasmussen, N. C. Nielsen, SIMPSON. A general simulation program for solid-state NMR spectroscopy J. Magn. Reson. 147, (2), 296 (2000).
[89]    J.-P. Kintzinger, H. Marsmann, Oxygen-17 and silicon-29, Springer-Verlag, Berlin, 1981.
[90]    J. Bill, T. W. Kamphowe, A. Müller, T. Wichmann, A. Zern, A. Jalowieki, J., Mayer, M. Weinmann, J. Schuhmacher, K. Müller, J. Peng, H. J. Seifert, F. Aldinger, Precursor-derived Si-(B-)C-N ceramics: thermolysis, amorphous state and crystallization Appl. Organomet. Chem. 15, 777 (2001)
[91]    G. D. Soraru, F. Babonneau, J. D. Mackenzie, Structural evolutions from polycarbosilane to silicon carbide ceramic J. Mater. Sci. 25, 3886 (1990)
[92]    Y. L. Li, E. Kroke, R. Riedel, C. Fasel, C. Gervais, F. Babonneau, Thermal cross-linking and pyrolytic conversion of poly(ureamethylvinyl)silazanes to silicon-based ceramics Appl. Organomet. Chem. 15, 820 (2001)
[93]    C. Gerardin, PhD thesis, University of Paris VI, 1991.
[94]    R. K. Harris, M. J. Leach, D. P. Thompson, Synthesis and magic-angle spinning nuclear magnetic resonance of nitrogen-15-enriched silicon nitrides Chem. Mater. 2, 320 (1990)
[95]    G. R. Hatfield, K. R. Carduner, Solid state NMR: applications in high performance ceramics J. Mater. Sci. 24, 4209 (1989)
[96]    E. Framery, PhD thesis, University of Rennes, 1996
[97]    C. Gervais, F. Babonneau, J. Maquet, C. Bonhomme, D. Massiot, E. Framery, M. Vaultier, N-15 cross-polarization using the inversion-recovery cross-polarization technique and B-11 magic angle spinning NMR studies of reference compounds containing B-N bonds Magn. Reson. Chem. 36, 407 (1998)
[98]    E. Brendler, E. Ebrecht, B. Thomas, G. Boden, T. Breuning, $^{15}N$ CP/MAS NMR as an instrument in structure investigations of organosilicon polymers Fresenius J. Anal. Chem. 363, 185 (1999)
[99]    H. Noeth, H. Vahrenkamp, Nuclear resonance investigations on boron compounds. I. $^{11}B$ nuclear resonance spectra of boranes with substituents from the first eight-membered period of the periodic system Chem. Ber. 99, 1049 (196)
[100]    H. Noeth, B. Wrackmeyer, NMR studies of boron compounds. VIII. Comparison of boron-11 and nitrogen-14 NMR data of tricoordinate boron compounds with carbon-13 and nitrogen-14 NMR data of carbonic acid derivatives and carbonium ions Chem. Ber. 107, 3089 (1974)
[101]    P.S. Marchetti, D. K. Kwon, W. R. Schmidt, L. V. Interrante, G.E. Maciel, High-Field B-11 Magic-Angle Spinning Nmr Characterization of Boron Nitrides Chem. Mater. 3, 482 (1991).
[102]    H. Noeth, W. Tinhof, B. Wrackmeyer, Nuclear magnetic resonance studies on boron compounds. VI. Nitrogen-14 and boron-11 NMR studies on silylamines and silylaminoboranes Chem. Ber. 107, 518 (1974)

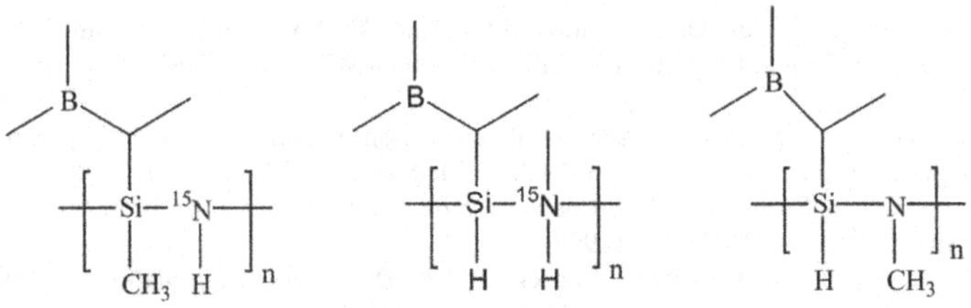

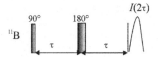

Figure 1. Chemical structures of precursor polymers **1**, **2** and **3**.

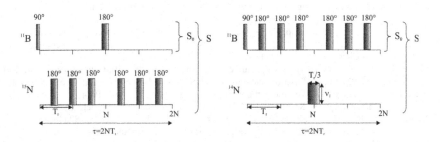

Figure 2. $^{11}$B Spin-echo pulse sequence.

Figure 3. $^{11}$B$\{^{15}$N$\}$ REDOR (left) and $^{11}$B$\{^{14}$N$\}$ REAPDOR (right) pulse sequences.

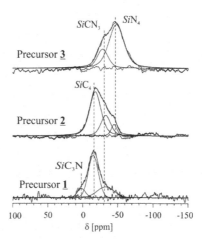

Figure 4. $^{29}$Si MAS NMR spectra of the pyrolysis intermediates at 1400 °C.

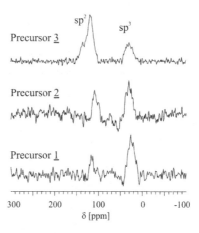

Figure 5. $^{13}$C MAS NMR spectra of the pyrolysis intermediates at 1400 °C.

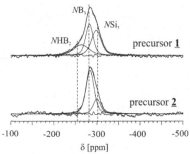

Figure 6. $^{15}$N MAS NMR spectra of the pyrolysis intermediates at 1400 °C.

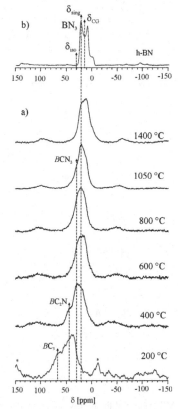

Figure 7. $^{11}$B MAS NMR spectra of a) precursor system **1** and b) pure h-BN. Dotted lines correspond to position of the low field singularity for tri-coordinated boron sites, $\delta_{sing}$.

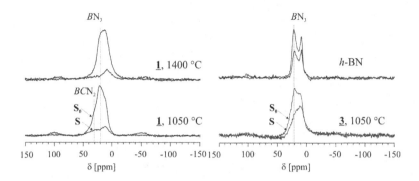

Figure 8. $^{11}B\{^{15}N\}$ REDOR (left) and $^{11}B\{^{14}N\}$ REAPDOR (right) spectra obtained after evolution time of $\tau = 1$ and $\tau = 0.6$ ms ($\nu_r = 10$ kHz). $S_0$ and S are reference and dephased spectra, respectively.

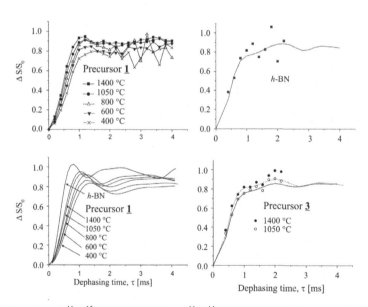

Figure 9. Experimental $^{11}B\{^{15}N\}$ REDOR (left) and $^{11}B\{^{14}N\}$ REAPDOR (right) curves. Solid lines are theoretical curves for a BN$_3$ spin system with the B-N distances given in Table 4.

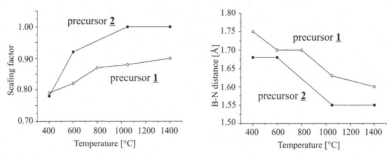

Figure 10. Scaling factors (left) and the B-N distances (right) against annealing temperature.

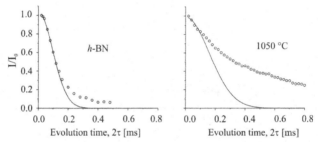

Figure 11. Experimental $^{11}$B spin-echo decay for h-BN and pyrolysis intermediates 1 at 1050 °C. Solid line corresponds to Gaussian line fitting with $2\tau < 0.1$ ms.

Table 1. $^{29}$Si chemical shift values observed in the present polymer-derived ceramics

| Molecular Unit | $^{29}$Si chemical shift [ppm] | Reference |
|---|---|---|
| SiC$_3$(sp$^3$)N | + 2 | 89 |
| SiC$_4$ | -15 | 90, 91 |
| SiCN$_3$ | -32 | 92, 93 |
| SiN$_4$ | -48 | 94, 95 |

Table 2. $^{15}$N chemical shift values observed in the present polymer-derived ceramics

| Molecular Unit | $^{15}$N chemical shift [ppm] | Reference |
|---|---|---|
| NHB$_2$ | -264 | 96, 97 |
| NB$_3$/NHBSi | -282 | 94, 98 |
| NSi$_3$ | -298 | 94, 98 |

Table 3. $^{11}$B chemical shift values for tri- coordinated BC$_x$N$_{3-x}$ units. $\delta_{sing}$ are chemical shift values for precursor system **1** and h-BN (low-field singularity, see text)

| Boron site | $\delta_{sing}$ [ppm] | Reference |
|---|---|---|
| NB$_3$ | - | 99,100 |
| h-BN | 22 | 101 |
| BCN$_2$ | 30 | 99 |
| BC$_2$N | 45 | 100,102 |
| BC$_3$ | 68 | 100 |

Table 4. $^{11}$B{$^{15}$N}REDOR and $^{11}$B{$^{14}$N} REAPDOR data.

| REDOR | Scaling factor | B-N distance [ Å] |
|---|---|---|
| **1**, 1400 °C | 0.90 | 1.60 |
| **1**, 1050 °C | 0.88 | 1.63 |
| **1**, 800 °C | 0.87 | 1.70 |
| **1**, 600 °C | 0.82 | 1.70 |
| **1**, 400 °C | 0.79 | 1.75 |
| **2**, 1400 °C | 1.00 | 1.55 |
| **2**, 1050 °C | 1.00 | 1.55 |
| **2**, 600 °C | 0.92 | 1.68 |
| **2**, 400 °C | 0.78 | 1.68 |
| REAPDOR | Scaling factor | B-N distance [ Å] |
| **3**, 1400 °C | 1.00 | 1.44 |
| **3**, 1050 °C | 1.00 | 1.55 |
| *h*-BN | 1.00 | 1.44 |

Table 5. $^{11}$B spin echo data for the h-BN and the pyrolysis intermediates.

| Precursor systems | B-B distance [ Å] |
|---|---|
| **1**, 1400 °C | 2.78 |
| **1**, 1050 °C | 2.98 |
| **2**, 1050 °C | 2.78 |
| **3**, 1400 °C | 2.85 |
| **3**, 1050 °C | 2.93 |
| *h*-BN | 2.52 |

# INTERMEDIATE-RANGE ORDER IN POLYMER-ROUTE Si-C-O FIBERS BY HIGH-ENERGY X-RAY DIFFRACTION AND REVERSE MONTE CARLO MODELLING

Kentaro Suzuya

Japan Atomic Energy Agency (JAEA, J-PARC), Tokai, Ibaraki 319-1195, Japan

Shinji Kohara

Japan Synchrotron Radiation Research Institute (JASRI, SPring-8), Hyogo 679-5198, Japan

Kiyohito Okamura

Japan Ultra-High Temperature Research Institute (JUTEM), Ube, Yamaguchi 755-0001, Japan

Hiroshi Ichikawa

Nippon Carbon Co., Ltd., Chuo-ku, Tokyo 104-0032, Japan

Kenji Suzuki

Advanced Institute of Materials Science (AIMS), Sendai 982-0252, Japan

## ABSTRACT

The atomic scale structure of polymer-derived Si-C-O ceramics fibers has been investigated by X-ray diffraction using high-energy synchrotron radiation at SPring-8 and reverse Monte Carlo (RMC) modelling. In the amorphous/microcrystalline Si-C-O fibers (NL400, heat-treated at 1000 ~ 1400 °C) the Si-atom prefers tetrahedral bonding to carbon and/or oxygen atoms, which leads to an inhomogeneous complex structure of SiC and Si-C-O domains and excess carbon region. By the RMC modelling, the basic structures of the Si-C-O fibers are the three-dimensionally connected $SiC_{4-x}O_x$ (x = 0, 1, 2, 3, and 4) tetrahedral random-network structure which is the successive assembly of the $SiC_{4-x}O_x$ tetrahedral unit which share elements (corners and edges) and the excess carbon atoms which are interconnected and distributed widely in the network cages.

## INTRODUCTION

Si-C-O fibers produced from inorganic precursors[1,2] have recently been intensively studied as potential candidates in the fabrication of thermo-structural composites for applications at high temperature.[3,4] In order to avoid the brittleness at grain boundaries of polycrystalline state, the molecular structure of the precursor and the pyrolysis process must be optimized in order to prevent crystallization and to preserve an amorphous state up the working temperature.[5,6] The retention of the amorphous state at higher temperatures appears to enhance the mechanical properties. It seems that the presence of Si-C-O phase increases the crystallization temperature of the fibers.[7]

Microstructural characterization of the Si-C-O fibers derived from polymer pyrolysis has been limited.[8,9,10,11] In fact, it has not been shown conclusively whether such materials are microcrystalline or amorphous. This paper describes the atomic-scale structure of the amorphous/microcrystalline Si-C-O fibers by high-energy X-ray diffraction[12] and reverse Monte Carlo modelling.

## EXPERIMENTAL PROCEDURE

The Si-C-O fiber samples used in the present study are industrial products NL400 of the Nippon Carbon Co. Ltd. The details of the material processing are given in several papers.[4] The NL400 fibers were produced from polycarbosilane (PC) of chemical composition 13.1 at% Si, 25.2 at% C, 0.1 at% O and 61.6 at% H and having a number-average molecular weight 2000. The PC was melt-spun into fine fibers, and then the fibers cured by thermal oxidation to make the Si-C-O fibers. In the thermal curing oxidation process, bundles of PCS fibers were heat-treated at 300 °C in air. NL400 fibers were obtained by heat-treatment of the cured PC fibers at 1000°C in Ar gas and having a composition of 36.6 at% Si, 44.5 at% C, 14.4 at% O and 4.5 at% H. Si-C-O fiber samples in this work were obtained by annealing of the NL400 fibers at temperatures of 1000, 1200 and 1400 °C in Ar gas for 0.5 h, respectively. Heat-treated PC powder as a Si-C standard sample was produced by heat-treatment of PC at 1000 °C in Ar gas for 0.5 h.

The high-energy X-ray diffraction (HEXRD) experiments have been performed in the transmission geometry using monochromatized X-rays of 61.7 keV ($\lambda$ = 0.2 Å) at BL04B2 2-axis diffractometer, SPring-8, Japan. The diffractometer was constructed for structural studies of amorphous materials which need a good real-space resolution to resolve the individual atom-atom correlations.[13] Structure factor, S(Q), was derived according to the Faber-Ziman definition from the scattered intensity measured up to Q (= $4\pi\sin\theta/\lambda$) = 22 Å$^{-1}$. This extended range of Q due to high-energy X-rays is essential for the better real-space resolution.

The fiber samples were arranged by bundling the fibers to form sheets about 0.1m thick and then arranged on an aluminum frame without enclosure. On the other hand, the powder sample was encapsulated into a very thin silica tube for the transmission measurements.

REVERSE MONTE CARLO MODELLING

The reverse Monte Carlo (RMC) modelling algorithm has been described in detail elsewhere,[14] in essence it is a computer simulation method for producing a set of particle configurations that is consistent with an input experimental structure factor S(Q). The resulting configurations can then be subjected to various geometrical analyses to reveal information about the short- and intermediate-range order of the complex disordered materials that cannot be obtained directly from the S(Q).

In RMC simulations, the difference between the experimental and calculated S(Q) is quantity to be minimize $\chi^2$ and a three dimensional configuration (structural model) consisted with the S(Q) is generated using the Monte Carlo algorithm which is a variation of metropolis method.

In this study, we used RMC++ program with 50000 particles (Si: 19160, C: 23300, O: 7540 and Si: 19060, C: 23400, O: 7540) in cubic cell (86 Å × 86 Å × 86 Å and 83.8 Å × 83.8 Å × 83.8 Å) which are estimated from the chemical compositions and atom number densities of the Si-C-O fiber samples (NL400 fibers, heat-treatment temperature HTT = 1000 and 1400 °C), respectively. For the RMC modelling, we used structure factors S(Q) of the heat-treated NL400 fibers (HTT = 1000 and 1400 °C) up to 20 Å$^{-1}$ obtained by high-energy X-ray diffraction experiments.

RESULTS AND DISCUSSION

The total structure factors S(Q) for three Si-C-O fiber samples (NL400, HTT = 1000, 1200 and 1400 °C) and the powder sample of heat-treated PC (HTT = 1000 °C) from HEXRD are shown in Fig. 1. The S(Q) for NL400 fiber (1000 °C) does not exhibit the weak diffraction lines due to the amorphous structure in the short-range atomic order. However the general aspect of the S(Q) for the NL400 fibers (1200 and 1400 °C) and the heat-treated PC sample exhibits similar profiles which show the diffraction lines of large amount of β-SiC microcrystal. The total correlation function T(r) obtained by Fourier transformation of the S(Q) from Fig. 1 are shown in Fig. 2 with the scaled T(r) of silica glass[15]. In the T(r) of NL400 fibers, two main peaks at around 1.9 and 3.1 Å are observed in the range of r < 3.5 Å. On the basis of the atomic sizes of the involved atomic species and the scattering weight W$_{i-j}$ (i-j: Si-C, Si-O,

Si-Si, C-O, C-C, O-O) those peaks can be attributed to individual atomic pairs. In the following discussion the total correlation functions T(r) of NL400 (1000 °C) fibers will be compared with those of the heat-treated NL400 fibers (1200 and 1400 °C), that of the heat-treated PC powder (1000 °C), and also with the structures of the related amorphous phases: SiC[16] and SiO$_2$ (silica glass)[17], and the related crystalline phases: β-SiC[17] and nanocrystalline graphite[18].

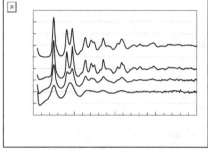

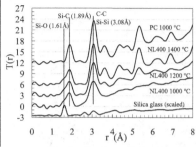

Fig. 1   Total structure factors                        Fig. 2   Total correlation functions

The small peak or shoulder at about 1.6 Å in the T(r) of NL400 fibers (HTT = 1000, 1200 and 1400 °C) belongs to Si-O correlation comparing to the scaled T(r) of silica glass[15]. With the conventional X-ray diffraction the Si-O peak is not observed separately due to the insufficient real-space resolution. In the study on silica glass,[15] the pair distribution function G(r) (= 4πrρT(r), ρ: atom number density) exhibits three peaks at 1.61 Å (Si-O), 2.63 Å (O-O) and 3.07 Å (Si-(O)-Si). The peak positions are good agreement with the distances in NL400 (1000 °C) at about 1.6 Å (shoulder), 2.6 Å (shoulder) and 3.1 Å (overlap). This suggests that there are silica-like tetrahedral SiO$_4$ units in the heat-treated NL400 fibers (1000 °C). On the other hand, the Si-O peak is not found in the T(r) of the heat-treated PC (1000 °C), because the PC sample did not through the oxidation curing process.

The main peak at around 1.9 Å reflects the Si-C correlation. The peak at around 3.1 Å in T(r) is mainly attributed to C-(Si)-C and Si-(C)-Si correlations where, however, also underlying contributions from other correlations are superposed. The positions of the two peaks are in good agreement with the corresponding distances in amorphous and crystalline SiC (Si-C at 1.88 ~ 1.89 Å, C-(Si)-C at 3.07 ~ 3.08 Å and Si-(C)-Si at 3.07 ~ 3.08 Å).[16,17]

The peaks of the heat-treated NL400 fibers are broadened on their left hand sides, compared with the peaks of PC (1000 °C) due to direct Si-O (1.6 Å) and indirect O-(Si)-O (2.6 Å) distances in a silica-like SiO$_4$ tetrahedral units as discussed above. Thus, assuming that the peak at 1.4 - 2.2 Å in the T(r) of the NL400 fibers contains two contributions, namely a contribution of a Si-O correlation at around 1.6 Å and a contribution of a Si-C correlation at around 1.9 Å, the corresponding partial average coordination number $N_{i-j}$ can be estimated from the peak area $A_{i-j}$ using $N_{i-j} = C_{i-j} \cdot A_{i-j}/W_{i-j}$. The areas $A_{i-j}$ were determined by fitting two Gaussian curves to the peak of the corresponding total radial distribution function RDF = $4\pi r^2 G(r)$ = rT(r). From the Gaussian at 1.61 Å, the average values $N_{Si-O}$ = 0.8 and $N_{O-Si}$ = 1.4, and from the Gaussian at 1.89 Å, the values $N_{Si-C}$ = 3.1 and $N_{C-Si}$ = 7.2 were obtained. The sum of $N_{Si}$ = $N_{Si-O}$ + $N_{Si-C}$ = 3.9 proves that the Si-atoms are tetrahedrally coordinated in the NL400 fibers. This strongly suggests not only the existence of SiO$_4$ and SiC$_4$ tetrahedra but also that of C-O mixed tetrahedral units Si(C,O)$_4$.

The presence of the Si-C-O phase or domain composed of SiC$_4$, SiO$_4$ and Si(C,O)$_4$ tetrahedra and of the excess free carbon atoms are important for understanding the origin of structural evolution of the fibers at high temperature. Thus, to study more detailed short- and intermediate-range ordering structures of the fibers we made three-dimensional structural model by RMC modelling with the HEXRD data. Figure 3 shows the configurations of Si-C and Si-O bonds in the RMC models for the heat-treated NL400 fibers (HTT = 1000 and 1400 °C). Both pictures clearly show the existence of $SiC_4$, $SiO_4$, and various $Si(C,O)_4$ ($SiC_3O$, $SiC_2O_2$, $SiCO_3$) tetrahedra and visibly suggests that these tetrahedra form network structure in which the tetrahedra are connected adjacent ones by corner- and edge-sharing.

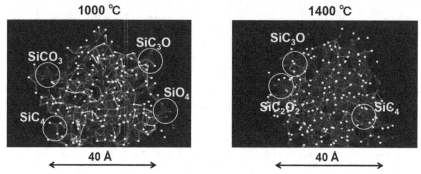

Fig. 3   Configurations of Si-C and Si-O bonds in RMC models for amorphous and microcrystalline Si-C-O fibers (NL400, HTT = 1000 and 1400 °C).

The tetrahedral network structure is commonly observed in both amorphous (NL400, 1000 °C) and microcrystalline (NL400, 1400 °C) Si-C-O fibers as the frame. Figure 4 is the schematic drawing of the Si-C-O network formed by $Si(C,O)_4$ tetrahedra in the amorphous and microcrystalline Si-C-O fibers. The tetrahedra are drawn as triangles schematically in the figure.

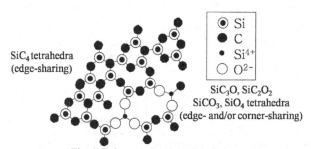

Fig. 4   Schematic RMC model of Si-C-O fibers.

The excess free carbon atoms (C-C bonds) are distributed in the cages of the Si-C-O network though are not drawn in Figs. 3 and 4. It is difficult to figure out the structure of the group of C-C bonds in the cages because the X-ray scattering weight $W_{C-C}$ of the Si-C-O fibers is the relatively smallest value giving very few structural information for C-C correlations.

These structural features in short- and intermediate-range order are commonly observed in both amorphous (NL400, 1000 °C) and microcrystalline (NL400, 1400 °C) Si-C-O fibers. No obvious structural effect of the crystallization is found between these two RMC models in the length- scale of Fig. 3. However, in the scale of 50000 atoms RMC model as shown in Fig. 5, the segregation of excess carbon atoms (C-C bonds) in the large-size Si-C-O network cages formed by crystallization is observed in the RMC model for the microcrystalline (NL400, 1400 °C) fibers.

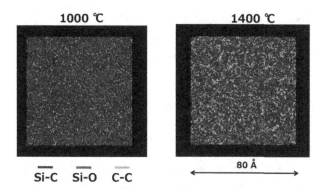

Fig. 5  Configurations of Si-C, Si-O, and C-C bonds in RMC models for amorphous and microcrystalline Si-C-O fibers (NL400, HTT = 1000 and 1400 °C).

The structural models of the amorphous/microcrystalline Si-C-O fibers proposed by HEXRD and RMC modelling are seemingly similar to, but actually different from, the models proposed in previous studies for Si-C-O ceramics.[10,12,19] The framework of the Si-C-O fibers is the three-dimensionally connected $SiC_{4-x}O_x$ (x = 0, 1, 2, 3, and 4) tetrahedral random-network structure which is the successive assembly of the $SiC_{4-x}O_x$ tetrahedral unit which share elements (corners and edges). However, no distinct area of $SiO_2$ is found though domain of SiC is observed in the RMC models. The excess carbon atoms are distributed in the network cages. More detailed structures of the Si-C-O tetrahedral network and the excess carbons of the Si-C-O fibers are under investigation by RMC modeling in combination with high-energy X-ray and neutron diffraction data.

CONCLUSION

The structures of polymer-derived Si-C-O fibers (heat-treated NL400) were investigated by high-energy X-ray diffraction and reverse Monte Carlo (RMC) modelling. In the amorphous/microcrystalline Si-C-O fibers (NL400, heat-treated at 1000 ~ 1400 °C) the Si-atom prefers tetrahedral bonding to carbon and/or oxygen atoms, which leads to an inhomogeneous complex structure of SiC and Si-C-O domains and excess carbon region. By the RMC modelling, the basic structures of the Si-C-O fibers are the three-dimensionally connected $SiC_{4-x}O_x$ (x = 0, 1, 2, 3, and 4) tetrahedral random-network structure and the excess carbon atoms which are interconnected and distributed widely in the network cages.

REFERENCES
[1]S. Yajima, J. Hayashi and M. Omori, *Chem. Lett.*, **9**, 931 (1975).

[2]S. Yajima, K. Okamura, T. Matsuzawa, Y. Hasegawa and T. Shishido, *Nature*, **279**, 706 (1979).

[3]A.R. Bunsell, and A. Piant, *J. Mater. Sci.*, **41**, 823 (2006).

[4]K. Okamura, T. Shimoo, K. Suzuya and K. Suzuki, *J. Ceram. Soc. Jpn.*, **114**, 445 (2006).

[5]T. Mah, D.E. McCullum, J.R. Hoenigman, H.M. Kim, A.P. Katz and H.A. Lipsitt, *J. Mater. Sci.*, **19**, 1191 (1984).

[6]G.D. Soraru, F. Babonneau, J.D. Mackenzie, *J. Mater. Sci.*, **25**, 3886, (1990).

[7]R. Bodet, N. Jia, and R.E. Tressler, *J. Eur. Ceram. Soc.*, **16**, 653 (1996).

[8]M. Monthioux, A. Oberlin and E. Bouillon, *Composite Sci. Tech.*, **37**, 25 (1990).

[9]E. Bouillon, D. Mocaer, J.F. Villeneuve, R. Pailler, R. Naslain, M. Monthioux, A. Oberlin, C. Guimon and G. Pfister, *J. Mater. Sci.*, **26**, 1517 (1991).

[10]C. Laffon, A.M. Flank, P. Lagarde, M. Laridjani, R. Hagege, P. Olry, J. Cotteret, J. Dixmier, J.L. Miquel, H. Hommel and A. P.Legran, *J. Mater. Sci.*, **24**, 1503 (1989).

[11]K. Suzuya, M. Furusaka, N. Watanabe, M. Osawa, K. Okamura, K. Shibata, T. Kamiyama and K. Suzuki, *J. Mater. Res.*, **11**, 1169 (1996).

[12]K. Okamura, K. Suzuya, S. Kohara, H. Ichikawa and K. Suzuki, *Key Engineering*, **352**, 65-68 (2007).

[13]S. Kohara, K. Suzuya, Y. Kashihara, N. Matsumoto, N. Umesaki and I. Sakai, *Nuc. Instr. Meth. Phys. Res.* A, **467-468**, 1030 (2001).

[14]R. L. McGreevy, *J. Phys.: Condens. Matter*, **13**, 877, (2001).

[15]S. Kohara and K. Suzuya, *J. Phys.: Condens. Matter*, **17**, S77, (2005).

[16]J.P. Rino, I. Ebbsjö, P.S. Branicio, R. K. Kalia, A. Nakano, F. Shimojo and P. Vashishta: *Phys. Rev.* B, **70**, 045207 (2004).

[17]P.T.B. Shaffer, *Acta Cryst.* B, **25**, 477 (1967).

[18]T. Fukunaga, K. Itoh, S. Orimo, M. Aoki and H. Fujii, *J. Alloys Comp.*, **327**, 224 (2001).

[19]A. Saha, R. Raj and D.L. Williamson, *J. Am. Ceram. Soc.*, **89**, 2188 (2006).

EVALUATION OF HEAT STABILITY OF Si-O-C FIBERS DERIVED FROM POLYMETHYLSILSESQUIOXANE

Masaki Narisawa, Ryu-Ichi Sumimoto, Ken-Ichiro Kita, Yayoi Satoh and Hiroshi Mabuchi
Department of Materials Science, Graduate School of Engineering, Osaka Prefecture University
1-1, Gakuen-Cho, Naka-Ku, Sakai 599-8531, Osaka, Japan
Young-Wook Kim
Department of Materials Science and Engineering, The University of Seoul, 90 Jeonnong-dong, Dongdaemoon-ku, Seoul 130-743, Korea
Masaki Sugimoto and Masahito Yoshikawa
Quantum Beam Science Directorate, Japan Atomic Energy Agency, Takasaki, Gunma, 370-1292, Japan

ABSTRACT
Si-O-C fibers were synthesized from melt spun polymethylsilsesquioxane (PMSQ) fiber with $SiCl_4$ curing. The cured and pyrolyzed fibers were investigated by FT-IR, an optical microscope and SEM and XRD. Heat resistance of each fiber was examined under various environments in an air flow, in an Ar flow with carbon black or in a $N_2$ flow with carbon black.

INTRODUCTION
Recently, we succeeded in synthesis of continuous Si-O-C ceramic fibers by using a kind of silicone polymer, polymethylsilsesquioxane (PMSQ), which is melt spinnable in a temperature range of 393-453K.[1] Since a softening point of PMSQ is relatively low, curing process, which is necessary for holding the fiber shape during the pyrolysis, must be performed up to 373K. It suggests that ordinary thermal oxidation curing available for carbon fiber or Si-C-O fiber production is not appropriate for such silicone fiber curing. Since the PMSQ fiber contains considerable amount of silanol groups even after melt spinning, cross-linking of silanol groups by Lewis acid is expected to be effective to cross-link component macromolecules.[2] Exposure to a vapor of $SiCl_4$, $TiCl_4$ or $BCl_3$ was found to be effective for holding fiber shape during pyrolysis, although the efficiency would correspond to acid strength of each agent and a vapor pressure in an adopted temperature. In practical, $SiCl_4$ curing is most promising for general use even in industry because of intrinsic low cost and easy handling. Heat resistivity of such Si-O-C fibers was, however, often claimed because it is intrinsically oxide base amorphous fiber, while

39

Si-C-O fibers derived from PCS possesses silicon carbide – like networks at least in a local range.[3] There is a promising report about oxidation resistance of such Si-O-C amorphous up to 1473K, which is accompanied with surface oxide layer formation in an oxidation atmosphere.[4] Viscoelastic nature and crystal growth in various Si-O-C materials were investigated and summarized from the viewpoint of fundamental science in recent years.[5,6] There are, however, no clues about practical heat resistivity of the Si-O-C fiber beyond 1473K in various environments, in which the fiber is expected to be used.

In this article, we describe synthesis of Si-O-C fibers with $SiCl_4$ curing method. Observation of the synthesized Si-O-C fibers heat-treated in various environments beyond 1473K is reported.

EXPERIMENT

Polymethylsilsesquioxane (YR 3370, Momentive Performance Materials Japan) in a form of a transparent solid was prepared for melt spinning process. The precursor was melt spun to fiber form at 423K. An averaged diameter of the spun fibers was 11.9 ±1.9 μm measured by Digital Microscope. For the chemical vapor curing, spun fibers was exposed to saturated vapor pressure of $SiCl_4$ in a closed enamel bat placed in a glove box at room temperature (297K). On the other hand, exposure of the spun fiber to $SiCl_4$ vapor in a glass tube was also examined (353 and 373K). The spun fiber and $SiCl_4$ in a Teflon dish were placed in a same glass tube with a determined distance. The heating rate of the tube was 40K/h. The spun and cured fibers were observed by an optical microscope with a transmitted light. IR spectra were obtained by a FT-IR spectrometer. After the pyrolysis at 1273K, morphology of the fibers was observed by FE-SEM. The obtained Si-O-C fibers were exposed to high temperature at 1511-1673K in a $MoSi_2$ furnace under various environments (air flow, Ar flow with carbon black (#3230B, Mitsubishi), $N_2$ flow with carbon black.

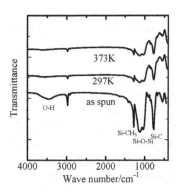

Fig. 1. IR spectra of spun or $SiCl_4$ cured fibers.

RESULTS AND DISCUSSION

The spun fiber was transparent, and there was no remarkable change about transparency and fiber appearance after curing. Mass gain after the $SiCl_4$ curing was 1.0, 6.2 or 7.5 mass% for 297, 373 or

393K curing. The ceramic yield was usually in a range of 84-86%. There was no marked influence of the adopted curing condition on the ceramic yield.

Figure 1 shows IR spectra of the spun or cured fibers. After the curing, absorption bands assigned to O-H absorption (3400cm⁻¹) has disappeared, while other absorption bands assigned to C-H (2900cm⁻¹), Si-CH₃ (1250cm⁻¹), Si-O-Si (1150-1100cm⁻¹) and (2900 cm⁻¹)and Si-C (820cm⁻¹) do not show remarkable change after the curing.

In a glance of the obtained fibers, the Si-O-C fiber synthesized at 373 or 393K is more flexible than that synthesized at 297K. Therefore, heat resistance of the fiber obtained at 373K is examined in the following study.

1 min exposure of the fiber at 1608K in air flow does not cause marked change on fiber morphology. 3h heat treatment at the same condition, however, causes formation of thick oxide layers on the fiber surface. Some area is fused to each other, and several cracks are observed on fiber bundles (Fig. 2 (a-b)). In some parts, irregularly thick oxide layer is formed (Fig. 2 (c-d)). Perhaps, CO gas evolution and resulting CO pressure caused debonding between fiber surface and silica layer. Thickness of such oxide layer is often fluctuated under this oxidation conditions. Sometimes, area of debonding between thick oxide layer and fiber surface area is observed (Fig. 2 (d)).

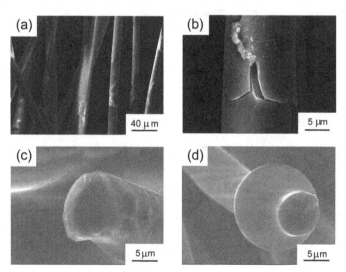

Fig. 2. SEM images of the Si-O-C fiber heat treated at 1608K for 3h in air; (a-b) Crack formation, (c-d) Fracture surfaces.

XRD patterns of the Si-O-C fibers after high temperature exposure in an air flow were summarized in Fig. 3 with references of quartz (pure silica) fibers heat-treated at the same conditions. XRD pattern of Si-O-C fiber after 1273K pyrolysis is similar to quartz fiber. The pattern is, however, broadened. After heat treatment at 1511K, Si-O-C fiber holds almost amorphous nature with a small peak of cristobalite at 22°, while the quartz fiber is completely crystallized. Even after heat treatment at 1608K, the XRD pattern of Si-O-C fiber holds partly amorphous nature as compared with the degraded quartz fiber. In a previous study, we have synthesized Si-O-C fiber with radiation curing, which has higher strength than the fiber obtained with $SiCl_4$ curing (tensile strength of $0.73\pm0.28$Pa for radiation curing and $0.30\pm0.13$GPa for $SiCl_4$ curing)[1]. The crystal growth in the Si-O-C fiber obtained with the radiation curing was somewhat rapid at the same heat treatment condition. Perhaps, captured surface oxygen played a role of nucleation sites.

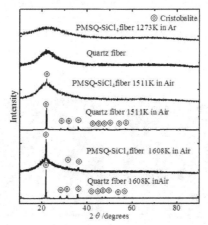

Fig. 3. XRD patterns of the Si-O-C fibers after heat treatment with reference patterns of quartz fibers.

For preparation of highly reduction atmosphere, the fiber bundles ($SiCl_4$ curing) were placed in a graphite crucible and carbon black was filled around the fiber. Under an Ar gas flow (1l/min), 3h heat treatment at 1608K was performed on the fiber. Fig. 4 shows resulting morphologies. The fiber is fragile and thin as compared with the fiber treated in air. The surface is often rough, and whiskers are formed at contact points between the monofilaments. Composition of the whiskers is probably SiC. Such whisker formation indicates SiO evolution and reaction with carbon sources in the vicinity of monofilaments.[7]

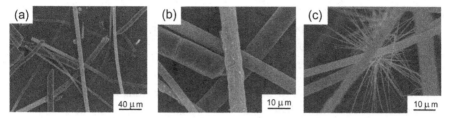

Fig. 4. SEM images of the Si-O-C fiber heat treated at 1608K for 3h in Argon with carbon black (CB); (a) Monofilaments, (b) Rough surface, (c) Whisker formation at contact points.

Fig. 5 shows the fiber morphology after heat treatment for 3h under 1673K in Ar or $N_2$ with carbon black. The fiber shape is partly maintained. Some parts are, however, broken to short cylindrical pieces. On residual fiber surface, silicon carbide tiny crystals are formed in an Ar atmosphere, and silicon nitride coarse crystals are formed on an $N_2$ atmosphere.

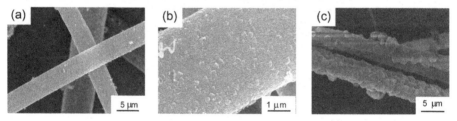

Fig. 5. SEM images of the Si-O-C fiber heat treated at 1673K for 3h; (a) Heat treated in Ar with CB, (b) Heat treated in Ar with CB (High magnification), (c) Heat treated in $N_2$ with CB.

CONCLUSION

Thin, continuous and flexible Si-O-C fibers are obtained from PMSQ with melt spinning and $SiCl_4$ vapor curing. In an oxidation atmosphere at high temperature, the fiber forms protective silica layers, which is thicken during a long exposure period. In XRD measurements, crystal growth (cristobalite) is apparently suppressed as compared with reference quartz fiber at the same oxidation conditions. In carbothermal reductive atmosphere, the fiber shows degradation at 1608-1673K with forming of SiC (in Ar) or $Si_3N_4$ (in $N_2$) crystals on surface.

ACKNOWLEDGEMENT

This work is partly supported by a Grant-in Aid for Scientific Research C (No. 20560627) from Japan Society of Promotion Science.

REFERENCES

[1] M. Narisawa, R. Sumimoto, K. Kita, H. Mabuchi, Y.-W. Kim, M. Sugimoto, and M. Yoshikawa, *Advanced Materials Research*, **66** 1-4 (2009).

[2] P. Baldus, M. Jansen, and D. Sporn, *Science*, **285**, 699-703 (1999).

[3] K. Suzuya, T. Kamiyama, T. Yamamura, K. Okamura, K. Suzuki, *J. Non. Crystal. Solids*, **150**, 167-171 (1992).

[4] C. M. Brewer, D. R. Bujalski, V. E. Parent, K. Su, and G. A. Zank, *J Sol-Gel Sci & Tech.*, **14**, 49-68 (1999).

[5] S. Modena, and G. D. Domenico Soraru, Y. Blum and R. Raji, *J. Am. Ceram. Soc.*, **88** 339-345 (2005).

[6] A. Scarmi, and G. D. Soraru and R. Raj, *J. Non-Crystalline Solid*, **351**, 2238-2243 (2005).

[7] W.-S. Seo, and K. Koumoto, *J. Am. Ceram. Soc.*, **79**, 1777-1782 (1996).

INVESTIGATION OF NANO POROUS SiC BASED FIBERS SYNTHESIZED BY PRECURSOR
METHOD

Ken'ichiro Kita, Masaki Narisawa, Atsushi Nakahira, and Hiroshi Mabuchi,
Graduate School of Engineering, Osaka Prefecture University
Sakai, Osaka 599-8531, Japan

Masayoshi Itoh
Fukushima National College of Technology
Iwaki, Fukushima 970-8034, Japan

Masaki Sugimoto, Masahito Yoshikawa
Quantum Beam Science Directorate, Japan Atomic Energy Agency
Takasaki, Gunma, 370-1292, Japan

ABSTRACT
        We synthesized SiC based ceramic fiber made from blend polymer of polycarbosilane and
polysiloxane. In the case of the polymer blends of PCS and polymethylphenylsiloxane (PMPhS), the
surface area of the ceramic fiber made from the polymer was high and showed the maximum of
$152m^2/g$, at 1673K. As for the polymer blends of PCS and polymethylhydrosiloxane (PMHS), the
maximum surface area of the ceramic fiber was $36.8 m^2/g$ at 1723K. The fiber, however, maintained
tensile strength of 0.35 GPa even after 1723K heat treatment.

INTRODUCTION
        Silicon carbide (SiC) based porous ceramics is one of promising thermoelectric materials,
thermal generators and gas separators in high temperature such as Ionide-Sulfer process [1]. In recent
years, there are many studies about Si-C base porous ceramics synthesized by polymer precursor
method [2,3]. There are, however, rare reports about porous ceramic fiber production available in a
mass scale in spite of its potential application.
        Recently, we have paid attention to various polysiloxanes for supplemental agents to
polycarbosilane (PCS) for Si-C-O ceramics fiber production [4]. The ceramic fiber derived from such
polymer blend often yields characteristic porous structure.
        In this study, we reported the microstructures of the fibers derived from various polymer
blends from viewpoint of fiber ceramization process.

EXPERIMENTAL PROCEDURE
        Commercialized polymethylphenylsiloxane (KF-54, Shin-Etsu Chemicals Co. Ltd., Japan)

and polymethylhydrosiloxanes (KF-99, Shin-etsu Chemicals Co., Ltd., Japan) without additional purification were blended with PCS (NIPUSI-Type A, Nippon Carbon, Japan). The blend ratios of polymethylphenylsiloxane (PMPhS) to PCS were 15 and 30 mass% and the blend ratios of polymethylhydrosiloxanes (PMHS) to PCS were 15 mass%. Polymer blend solutions in benzene were freeze-dried and white powders were obtained.

These polymer blends were melt-spun into pre-ceramic fibers at 523-578K. The prepared fibers including PMPhS were identified as PSxx (xx is the mass of contained PMPhS) and the fiber including 15mass% of PMHS were identified as HS15.

PS15 and PS30 were cured by γ-ray in air or electron beam in He atmosphere. The dose rate of γ-ray was fixed to $6.45 \times 10^2$ C/kg·h and the irradiation time was 96h. The PS15 and PS30 cured by γ-ray were identified as PS15-G and PS30-G. As for the electron beam curing, the first dose rate was 0.4kGy/s and the fibers were irradiated at 1000s, the second dose rate was 0.79kGy/s and irradiation time to the fiber was 2000s after the first irradiation, the final dose rate was 1.58kGy/s and irradiation time was 4000s after the second irradiation, and the fibers were heated at 773K in order to prevent the fibers from oxidizing by excess electron on the fibers [5]. The PS15 and PS30 cured by electron beam were identified as PS15-E and PS30-E. On the other hand, the HS15 were cured by thermal under flowing air and heating rate was 8 or 200K/h up to 485K. The HS15 cured by heating rate of 8K/h was identified as HS15-8K and the HS15 cured by heating rate of 200K/h was identified as HS15-200K. After the curing, all the fibers were pyrolyzed at 1273K for 1h in inert atmosphere. After the pyrolysis, the pyrolyzed fibers were heated at 1473-1773K for 30minutes in flowing Ar and the ceramics residue and the surface area of all the fibers was investigated.

RESULT AND DISCUSSION

Figure 1 shows that the mass residues and ceramics yields of PS15s, PS30s and HS15s after curing and pyrolysis at 1273K. The mass residues and ceramics yields of the fibers after melt-spinning were adopted as each standards, and the following data of the mass residues and ceramics yields of the fibers after curing and pyrolysis were the comparisons to the each standards. The mass gains of PS15-G and PS30-G after curing were 9.67% and 7.48%, and the mass gains of HS15-8K and HS15-200K after curing were 5.80% and 4.14%. These mass gains correspond to the oxidation which makes the Si-O-Si and Si-O-C cross-linking [6]. On the other hand, the mass gains of PS15-E and PS30-E were -2.51% and -5.92%. These mass losses were brought by annealing after irradiation of electron beam.

Figure 2 shows the residual masses of PS15s, PS30s and HS15s at 1273-1773K. The ceramics yields of PS15-G and PS30-G decrease rapidly at 1573-1673K. Those of PS15-E and PS30-E also decrease mainly at 1573-1673K, however, mass losses at 1273-1773K were only around 5%. As for HS15-8K and HS15-200K, the residual masses mainly decrease at 1573-1723K.

Figure 3 shows the surface areas of the fibers after 1573-1773K heat treatments. The surface area of PS15-G started to increase at around 1573K and that of PS15-G shows the maximum of $102m^2/g$ at around 1673K. The tendency of increasing of surface area of PS30-G in each temperature was the same as that of PS15-G, however, the value of the surface area of PS30-G was 1.5times of that of PS15-G in the same pyrolysis temperature. As for PS15-E and PS30-E, the surface areas of the fibers after pyrolysis at any temperature were not measured because the values were less than measurement limit value of our measuring device. In the case of HS15-8K, the surface area of the fiber started to increase at around 1673K and shows the maximum of $36.8m^2/g$ at around 1723K. As for HS15-200K, the surface area at 1723K was $13.0m^2/g$.

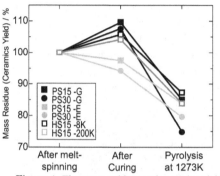

Figure 1. The mass residues and ceramics yields of PS15s, PS30s and HS15s after curing and pyrolysis at 1273K.

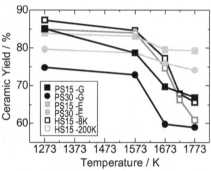

Figure 2. The ceramics yields of PS15s, PS30s and HS15s after pyrolysis at 1273-1773K.

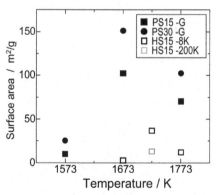

Figure 3. The surface areas of the fibers after pyrolysis after 1573-1773K .

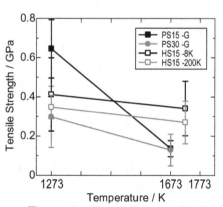

Figure 4. The tensile strength after pyrolysis 1273-1773K of PS15-G, PS30-G, HS15-8K and HS15-200K.

Figure 4 shows the tensile strength of PS15-G, PS30-G, HS15-8K and HS15-200K after pyrolysis 1273, 1673 and1773K which are suitable temperature to obtain porous ceramics fibers. The tensile strengths of all the fibers pyrolyzed at 1273K were less than 0.7GPa and they were weak as compared with Si-C-O fibers such as Nicalon(R). The causes of these tensile strengths were caused by hollows in the fibers derived from PMHS [4] and cracks on surface of the fibers derived from PMPhS.

Figure 5 shows the FE-SEM images of the cross sections of PS15-Gs pyrolyzed at 1573, 1673, 1773K and PS15-E pyorlyzed at 1673K. The cross section of PS15-G pyrolyzed at 1573K could not be observed, that of PS15-G pyrolyzed at 1673K showed pores whose average diameter was less than 100nm and that of PS15-G pyrolyzed at 1773K shows pores whose average diameter was among 100nm. The PS15-G pyrolyzed at 1773K seems to be sintered. On the other hand, the cross sections of PS15-E pyrolyzed at 1673K do not show any indication of pore formation.

Figure 6 shows the FE-SEM images of the cross sections of HS15-8K and HS15-200K pyrolyzed at 1723K, which contain large pores at the center of the fiber. Though the cross section of HS15-8K and HS15-200K are similar to PS15-G pyrolyzed at 1673K and 1773K, the surface areas of HS15-8K and HS15-200K were not beyond that of PS15-G. Structure may be more sintered than that of PS15-G during heat treatment.

Figure 5.   SEM images of the cross sections of PS15-Gs pyrolyzed at 1573, 1673, 1773K and PS15-E pyrolyzed at 1673K.

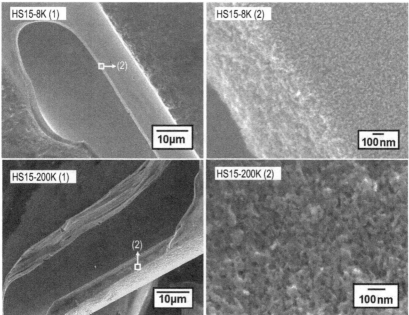

Figure 6.    SEM images of the cross sections of HS15-8K and HS15-200K pyrolyzed at 1723K.

The residual masses of sample fibers were rapidly decreased between 1573K and 1673K except PS15-E and PS30-E (Figure 2). Since it is well-known that CO gas is actively evolved from cured Si-C-O polymer in the same temperature range, it is accepted that the main cause of the mass loss of the fibers in the temperature range is CO gas evolution [7]. Since the surface area of these fibers were rapidly gained (Figure 3), it is conceivable that the increase of surface areas has deep relationship with the gas evolution. In the case of PS15-E and PS30-E, oxygen in curing was not supplied to the fibers, and there were no remarkable pores and mass loss in spite of heating at 1673K which is the best temperature for making porous ceramics fiber from PS15-G. From the above result, it is predicted that the excess oxygen captured by curing in the fibers is related to the increase of surface area.

In spite of the amount of fine pores in the fibers and 1673K or 1723K pyrolysis, PS15-G and PS30-G keep around 0.15GPa of tensile strength, and HS15-8K and HS15-200K keep around 0.3GPa of tensile strength (Figure 4). Generally, SiC crystals are going to grow up in the Si-C-O based fiber while heating beyond 1573K and the fiber is going to break to pieces because of grain boundaries sliding among the grown SiC crystals after heating at beyond 1773K [8]. The size of SiC crystal in Si-C-O based fiber after heating at 1773K is more than 1μm, however, the size of the crystals in PS15-G heated at 1773K (Figure 5) is about 100nm. It is suspected that the component of polysiloxane prevents the crystal from growing up and binds it together. Besides, the size of the crystals in HS15-8K and HS15-200K heated at 1723K looks like equal or smaller than that of PS15-G heated at 1673K, however, the surface areas of the HS15s were smaller than that of the PS15-G. It is suggested that pores are only existed on the surface of the HS15s. The diffusion speed of oxygen in such fibers by thermal oxidation curing is fast enough to oxidize uniformly all over the fibers, therefore it is possible that thermal resistance is brought by PMHS and the slight impurities in inert gas make the pores.

CONCLUSION

We synthesized a SiC based ceramic fiber made from the precursor polymer containing PCS and polysiloxane. In the case of the polymer blend containing PCS and PMPhS, γ-ray oxidation curing and pyrolysis at 1673K was the best condition to achieve large surface area and the polymer blend with more PMPhS was appropriate to increase the specific surface area. As for the polymer blend containing PCS and PMHS, the surface area of the ceramic fiber made from the polymer blend was less than that of the ceramic fiber made from the polymer blend of PMPhS, however, the average tensile strength of the ceramic fiber after pyrolysis at 1723K was around 0.35GPa. It seems that the component of polysiloxane plays 2 roles of a deterrent to SiC crystal growth and a binder among the crystals in the porous ceramic fiber.

REFERENCE

[1]M. Nomura, A. Kasahara, H. Okuda, and A. Nakao, Evaluation of the IS process featuring membrane techniques by total thermal efficiency, *Int. J. Hydrogen Energy*, **30**, 1465-73 (2005).
[2]P. Colombo, and M. Modesti, Silicon Oxycarbide Ceramic Foams from a Preceramic Polymer, *J. Am. Ceram. Soc.*, **82**, 573–78 (1999).
[3]Y. -W. Kim. K. -H. Lee, S. -H. Lee, and B. Park, Fabrication of Porous Silicon Oxycarbide Ceramics by Foaming Polymer Liquid and Compression Molding, *J. Ceram. Soc. Jpn.*, **11**, 863-4 (2003)
[4]K. Kita, M. Narisawa, H. Mabuchi, M. Itoh, M. Sugimoto, and M. Yoshikawa, Formation of continuous pore structures in Si-C-O fibers by adjusting melt spinning condition of polycarbosilane - polysiloxane polymer blend, *J. Am. Ceram. Soc.*, **92**, 1192-7 (2009).
[5]M. Sugimoto, A. Idesaki, S. Tanaka, and K. Okamura, Development of silicon carbide micro-tube from precursor polymer by radiation oxidation *Key Eng. Mater.*, **247**, 133-6 (2003).
[6]T. Shimoo, F. Toyoda and K. Okamura, Thermal stability of low-oxygen silicon carbide fiber (Hi-Nicalon) subjected to selected oxidation treatment, *J. Am. Ceram. Soc.* **83**, 1450-6 (2000).
[7]E. Caberry, and R. West, Decamethylcyclopentasilane and tetradecamethylcycloheptasilene, *J.Organometal. Chem.*, **6**, 582 (1966).
[8]T. Shimoo, H. Chen and K. Okamura, High-temperature stability of Nicalon under Ar or $O_2$ atmosphere, *J.Mater.Sci.*, **29**, 456-63 (1994).

# Processing
# and Applications

# MULLITE MONOLITHS, COATINGS AND COMPOSITES FROM A PRECERAMIC POLYMER CONTAINING ALUMINA NANO-SIZED PARTICLES

E. Bernardo[♠,*], G. Parcianello, P. Colombo[&,*]
University of Padova, Dipartimento di Ingegneria Meccanica – Settore Materiali
via Marzolo, 9, 35131 Padova, Italy

J. Adler, D. Boettge
Fraunhofer Institut Keramische Technologien und Systeme IKTS, Winterbergstrasse 28, 01277 Dresden, Germany

ABSTRACT
    Pressed silicone resin powders containing alumina nano-particles enabled the production of dense and crack-free mullite monoliths upon heat treatment at 1250°C in air. The chemical interaction between the preceramic polymer and the nano-filler determined a low energy barrier of nucleation (<700 kJ/mol) for the nucleation of mullite, with a high yield (about 80 vol%, after 100 s at 1350°C). The composite powders were obtained by dissolving the silicone resin in a solvent and dispersing the nano-sized alumina filler particles, followed by drying. The drying of the suspensions directly on a substrate, such as Si-SiC monoliths or SiC open-celled foams, allowed the preparation of thin and strongly bonded anti-oxidation coatings (tested in air at 1300°C). With the introduction of secondary micro- and nano-powders (mullite, yttria-stabilized or non-stabilized zirconia) in the suspensions, tetragonal zirconia-reinforced mullite matrix composites and thick mullite coatings on Si-SiC were obtained, after treatment at 1350°C for 1-3 h.

INTRODUCTION
    Mullite is well known to possess outstanding physical properties, such as low theoretical density (about 3.2 $g/cm^3$), low thermal conductivity (2 $W \times m^{-1} \times K^{-1}$), low thermal expansion ($\alpha_{20°/1000°C}$ from 4 to $5 \times 10^{-6}$ $K^{-1}$), low dielectric constant (6.5 at 1 MHz), and excellent creep and thermal shock resistance. These characteristics make mullite an excellent candidate for advanced ceramic applications such as gas filters, heat exchangers, multilayer packaging and window material in the mid-infrared range. It is also used in ceramic fibers and matrices of ceramic matrix composites to be used for furnace burners, catalytic converter substrates and for thermal protection systems for combustors in gas turbine engines.[1-3] A variety of crystallization reactions from synthetic precursors leading to the formation of mullite ceramics of high chemical purity, high sinterability and low mullitization temperatures have been reported.[3] Mullite precursors can be classified as either single phase, or diphasic, as a result of the starting materials and of the synthesis conditions employed.[4] Single phase precursors exhibit direct mullitization from the amorphous state at temperatures as low as ~950°C, while diphasic precursors mullitize above 1200°C by the reaction of transient spinel-type alumina with silica.
    A novel synthesis method, based on γ-alumina nanoparticles and a silicone resin, has been recently proposed by Bernardo et al.[5] This approach which consists of silicone powders mixed with nano-sized fillers can be extended also to the production of other ceramic systems (such as SiAlON and wollastonite ceramics),[6-7] and has several technological advantages over sol-gel-based precursors (the ones which generally yield the best purity and the lowest mullitization temperatures). In fact,

---

[♠]Corresponding author: enrico.bernardo@unipd.it
[*]Member, American Ceramic Society
[&]and Department of Materials Science and Engineering, The Pennsylvania State University, University Park, PA 16802

preceramic polymers can be easily shaped using conventional plastic-forming technologies, don't have any drying problems which hamper the possibility of fabricating bulk components and don't require any specialized handling procedures. In addition, using this approach the produced mullite is phase-pure even at low firing temperatures, contrary to the results from other researchers, who studied silicone resins filled with much larger $\alpha$-Al$_2$O$_3$ particles and/or metallic Al powder.[8,9] Griggio et al.[10] demonstrated that this novel method features also favorable kinetic properties, such as a low energy barrier of nucleation (<700 kJ/mol) and a high mullite yield (about 80 vol%, after 100 s at 1350°C). In the present work we extended the study, by introducing some processing modifications, to the preparation of zirconia-toughened mullite monoliths and anti-oxidation coatings on SiC-based ceramics. The second application is intended to provide a new method for producing oxide coatings (in general more resistant than silicon-based ceramics to corrosive environments: silicon-based ceramics exhibit excellent oxidation resistance in clean, dry oxygen, by forming a slow-growing, dense silica scale, but this scale can be severely degraded by reacting with impurities, such as alkali salts or water vapor). Mullite is a good candidate for coating Si-based ceramics mainly because of its low coefficient of thermal expansion, close to that of SiC; so far, whoever, the most effective coatings have been produced by expensive processes, such as CVD and plasma spray.[11]

EXPERIMENTAL

a) Preparation of mullite-based monoliths

The preceramic polymer consisted of a polymethylsiloxane (MK, Wacker-Chemie GmbH, Munchen, Germany) in powder form. It was dissolved in acetone under magnetic stirring for 15 minutes (3.2 g silicone resin for 100 mL acetone), thus producing a solution with 3.2% solid content. $\gamma$-Al$_2$O$_3$ nanopowders ("Aluminium oxide C," Degussa, Hanau, Germany; 15 nm mean particle size, specific surface area of 100 m$^2$/g) were added to the solution, in the weight ratio silicone/$\gamma$-Al$_2$O$_3$ = 1/2.125, again under magnetic stirring. This ratio was suggested by considering the ceramic yield of the polymer after heating in air (about 85 wt%) and the stoichiometry of mullite. The mixture was ultrasonicated for 10 min, producing a stable and homogeneous dispersion of alumina nanoparticles, in which no sedimentation was observed. The dispersion was poured into a glass container and dried at 60°C overnight. After evaporation of the solvent, a solid silicone-alumina nanocomposite mixture was obtained, in which the nanosized filler was homogeneously distributed. The mixture was finely ground (up to about 100 μm) and subsequently cold pressed at 40 MPa in a cylindrical steel die. The pressed samples (diameter ~31 mm, height ~1.5 mm) were inserted in a preheated furnace and heat-treated in air at 1350°C for 1 h, with a heating rate of 10°C/min. Mullite matrix composites were developed by introducing secondary powders in the silicone/$\gamma$-Al$_2$O$_3$/acetone suspensions and repeating the same procedures used for mullite monoliths. The secondary powders consisted of zirconia nanopowders (TZ-0, Tosoh Corporation, Tokyo, Japan, 40 nm diameter, 14 ± 3 m$^2$/g specific surface area; VP PH, Degussa, Hanau, Germany, 13 nm diameter, 100 m$^2$/g specific surface area; VP 3YSZ - doped with 3% mol. Y$_2$O$_3$, Degussa, Hanau, Germany, 13 nm diameter, 100 m$^2$/g specific surface area), and they were introduced in an amount corresponding to 20 wt% of the final ceramic.

b) Mullite coatings on SiC-based ceramics

The first experiments were performed with the above described suspensions of alumina nanopowders in an acetone solution of MK polymer (formulation A). Alternative water-based formulations (formulations B-D), reported in Table I, were used for further experiments; these formulations contained also a non-ionic surfactant (Pluronic P65, BASF Corporation, Florham Park, NJ) as dispersant aid. The composite powders, used for formulations C and D, were those previously used for mullite monoliths; formulation D included pre-formed mullite micro-particles (3-5 μm, Symulox M72 MC, Nabaltec AG, Schwandorf, Germany) added to reduce the shrinkage during sintering. For all the formulations the components were put in water under magnetic stirring for 10-15 min, followed by sonication for 10 min. Si-SiC monoliths (produced and provided by IKTS) were dip-coated with the above described acetone- or water-based dispersions. The coating speed was 6 cm/min; the coated sample were dried at 60°C for 1 h. SiC open-celled foams (produced and provided by IKTS - with different pore density, varying from 10 to 30 ppi -, in the form of blocks with dimensions about 25 × 10 × 10 mm, cut from larger samples) were coated by applying two or three cycles of dipping/centrifugation (at 1000 rpm) for 1 min/drying at 60°C for 1 h.

All the coating procedures led to deposits of silicone resin mixed with fillers, which were transformed into mullite layers by heat treatment at 1350°C for 3 h, with a heating rate of 10°C/min.

Table I. Composition of water-based formulations for mullite coatings on Si-SiC.

| Formulation | Components | | | | | | |
|---|---|---|---|---|---|---|---|
| | Distilled water | $\gamma$-$Al_2O_3$ | MK | MK solution | Composite powders | Mullite powders | Surfactant |
| B | 30 cc | 6 g | 2.82 g | | | | 1 g |
| C | 30 cc | | | | 7 g | | 0.5 g |
| D | 30 cc | | | | 5 g | 5 g | 1 g |

c) Sample characterization

Powdered monoliths were investigated by X-ray diffraction (XRD; Bruker D8 Advance, Karlsruhe, Germany), using CuKa radiation ($\lambda$=0.15418 nm). Data were collected between 15° and 62° $2\theta$ in step scan mode with step of 0.05° and a counting time of 5s/step in a Bragg-Brentano configuration. The XRD patterns were refined, according to the Rietveld's method, using the MAUD (material analysis using diffraction) program package.[12] This software allows the identification of the phases presents together with the quantification of their volume and of the dimension of the crystallites. Mullite and mullite-zirconia monoliths were subjected to indentation analysis, for microhardness and indentation fracture toughness determinations, operating with a load of 10 N and 100 N, respectively. The indentation fracture toughness was estimated using the method suggested by Anstis et al,[13] assuming an elastic modulus of 227 GPa (a value for the elastic modulus of mullite ceramics which is reported in the literature)[3] for both unreinforced mullite and mullite-zirconia composites. The microstructure was characterized by scanning electron microscopy (JSM-6490, JEOL, Tokyo, Japan); the residual porosity was measured by gas pycnometry (Micromeritics AccuPyc 1330, Norcross, GA).

The characterization of coatings was conducted by means of SEM and X-ray diffraction (for Si-SiC monoliths), with data collected between 15° and 75° $2\theta$ in step scan mode with step of 0.05° and a counting time of 3s/step in a thin film configuration (0.5° glancing angle). Oxidation tests were performed in oxidizing atmosphere (static air) at 1300°C for 100 h. After oxidation treatment, samples were finely ground and the oxidation resistance was quantified investigated using X-ray diffraction, analyzing the intensity of the peak at $2\theta$=21.9°, associated to the cristobalite that forms during SiC oxidation.

RESULTS AND DISCUSSION

a) Mullite-based monoliths

A recent publication was devoted to the study of the kinetics of mullite formation from MK silicone filled with $\gamma$-$Al_2O_3$ nanopowders:[10] the main findings were that the reaction between the preceramic polymer and the nano-filler determined a low energy barrier of nucleation (<700 kJ/mol), and a high mullite yield (about 80 vol%. after only 100 s at 1350°C). This temperature was chosen as a reference for the present investigation; a soaking time of 1 h was selected in order to achieve a virtually complete formation of mullite. Table II reports that this thermal treatment yielded samples possessing a density of 3.16 g/cm$^3$, corresponding to 98.8% of the theoretical density of mullite, with an average crystal size of about 180 nm, estimated by Rietveld refinements. The introduction of zirconia nanoparticles was significant, first of all, for the fact that the secondary filler did not affect the mullite formation, and we can therefore exclude any major Al$_2$O$_3$-ZrO$_2$ or polymer-ZrO$_2$ interaction during heating (for instance to give zircon). Table II indicates that the size of mullite crystals did not show any significant variation; the size of the zirconia grains did not exceed 110 nm. Fig. 1 shows that the most significant transformation in the polymer-alumina-zirconia mixtures occurred in the phase assemblage of the zirconia particles. In fact, Fig. 1a shows that the different zirconia powders, in the as-received state, were either mostly monoclinic (Tosoh TZ-0), monoclinic-tetragonal (Degussa VP PH) or completely tetragonal, if yttrium doped (Degussa VP 3YSZ). In all composites, see Fig. 1b, the amount of tetragonal phase was largely dominant. For the pure zirconia powders (TZ-0 and VP PH) this fact is likely due to the mechanical constraint operated by the mullite matrix on the zirconia particles upon thermal treatment. At the maximum heating temperature (1350°C) zirconia was undoubtedly tetragonal, and the tetragonal-monoclinic martensitic transformation was prevented by the retention of the crystal size well below the critical value (the zirconia particles were well below the critical size, which is in the order of 500 nm).[14] This can be attributed to the fact that the zirconia powder was homogeneously dispersed within the preceramic polymer (the starting nanoparticles had only a limited agglomeration).

The retention of a large amount of tetragonal phase is profitable for the mechanical properties, since stress-induced transformation may be exploited. Although a complete mechanical characterization is still in progress, preliminary indentation tests showed an almost three-fold increase of fracture toughness, from 2.7 ± 0.4 MPa m$^{0.5}$, for unreinforced mullite, to 7.2 ± 0.7 MPa m$^{0.5}$, for mullite reinforced with VP PH ZrO$_2$ (the increase is significant also for the fact that we used the elastic modulus for mullite also for the computation of toughness for the composite, determining an underestimation of the fracture toughness value). Schneider and Komarneni[3] reported that the toughness of analogous composites hardly exceeds 5 MPa m$^{0.5}$, and about 7 MPa m$^{0.5}$ may be achieved only with a substantial addition of other reinforcements (e.g. 40% vol., SiC whiskers). Fig. 1c illustrates that stress-induced transformation was effective for composites containing pure zirconia, since a substantial amount of monoclinic phase appeared after they were ground into fine powders (below 100 μm).

Table II. Summary of physical and mechanical properties of the investigated mullite-based monoliths.
M = mullite; Zt = tetragonal ZrO$_2$; Zm = monoclinic ZrO$_2$

| Composition | Bulk density (g/cm$^3$) | % of theoretical density | Crystal size (nm) | Phase assemblage (vol%) | Hardness (GPa) | Indentation $K_{IC}$ (MPa m$^{0.5}$) |
|---|---|---|---|---|---|---|
| Pure mullite | 3.16 ± 0.01 | 98.8 | M 181 ± 30 | M ~100% | 7.4 ± 0.5 | 2.7 ± 0.4 |
| Mullite - 20% ZrO$_2$ TZ-0 | 3.70 ± 0.02 | 99.2 | M 153 ± 30 Zt 108 ± 30 Zm 69 ± 30 | M 81.3 Zt 14.0 Zm 4.7 | 8.9 ± 0.6 | 6.6 ± 0.5 |
| Mullite - 20% ZrO$_2$ VP PH | 3.72 ± 0.02 | 99.7 | M 168 ± 30 Zt 102 ± 30 Zm 68 ± 30 | M 80.6 Zt 18.2 Zm 1.2 | 8.9 ± 0.5 | 7.2 ± 0.7 |
| Mullite - 20% ZrO$_2$ VP 3YSZ | 3.71 ± 0.02 | 99.5 | M 175 ± 30 Zt 103 ± 30 Zm - | M 81.3 Zt 18.7 Zm 0 | 13.5 ± 0.4 | 4.4 ± 0.8 |

The composite containing yttria-stabilized zirconia (VP 3YSZ) remained instead practically unchanged after grinding. The absence of stress-induced transformation is consistent with the observations by Becher et al.:[15] yttrium has a great effect in depressing the temperature of martensitic tetragonal-monoclinic transformation, which also decreases with decreasing size of zirconia. Therefore, nano-sized yttria-stabilized zirconia (VP 3YSZ), as-received or embedded in mullite, was consequently more stable in the tetragonal form. On the contrary, the retention of tetragonal phase in nano-sized pure zirconia (VP PH, or TZ-0) was purely due to a mechanical effect, so that the particles transform more easily to monoclinic zirconia during the propagation of cracks. For the composite containing yttria-stabilized zirconia, the limited increase of fracture toughness could be due to the occurrence of crack deflection and crack bowing at zirconia grains, known as secondary toughening effects.[3] Fig. 2 confirms the fact that the prepared ceramics were fine grained, with most crystals below 200 nm.

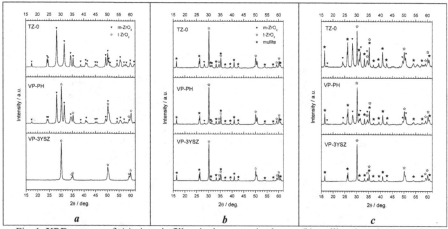

Fig. 1. XRD patterns of: (a) zirconia fillers in the as-received state; (b) mullite-zirconia composites after firing; (c) mullite-zirconia composites after grinding.

Fig. 2. SEM micrograph of tetragonal zirconia-reinforced mullite: (a) TZ-0; (b) VP PH; (c) VP-3YSZ.

b) Mullite coatings on SiC-based ceramics

The coatings were firstly made by dipping Si-SiC monoliths directly into the dispersion of alumina nanoparticles in an acetone solution of MK polymer. This operation resulted in an homogeneous layer, which was successively transformed into a mullite coating by heat treatment, as demonstrated by Fig. 3a (A coating on substrate). The fact that XRD patterns shows Si and SiC peaks, besides those of mullite layer, is an evidence of the limited thickness of the layer (about 5 μm, from SEM analysis); the formation of cristobalite is consistent with the fact that the coating was microcracked, thus leaving the substrate partially uncoated and able to react with atmosphere during heating. The microcracking was mainly due to the large shrinkage associated with the mullitization reaction occurring between the polymer and the filler (about 20% linear).[5]

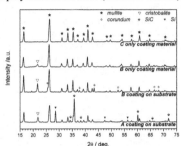

Fig. 3. Phase evolution in selected coatings deposited on Si-SiC monoliths.

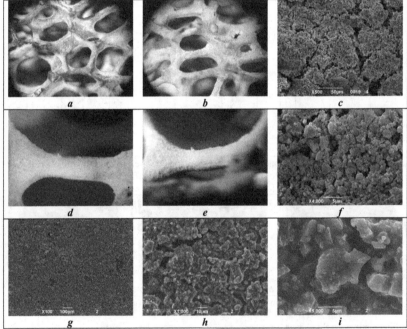

Fig. 4. Optical and SEM micrographs of selected coatings on SiC foams (two coating cycles, 10 ppi): a,b,c = Formulation C; c,d,e = Formulation D; g,h,i = Formulation D, sintering at 1400°C.

The water-based mixtures were prepared in order to provide a thick coating, with a simple procedure, not employing flammable acetone. For formulation B, alumina nanopowders and MK polymer powders (with a dimension of about 100 μm) were deposited separately: the melting of the silicone, upon the initial heating, caused a quite effective *in-situ* mixing with alumina, which resulted in mullite formation, as demonstrated by Fig. 3a (B coating on substrate). Although still microcracked, the coating were thicker (about 20 μm), and there is no evidence of peaks from the substrate. It must be noted that the presence of cristobalite in this case was not simply due to the oxidation of the substrate; in fact, some cristobalite formed even by treating only the coating material (Fig. 3a – B only coating material). The cristobalite formation, in this last case, can be reasonably attributed to the absence of quasi-molecular mixing between the silica source (MK polymer) and alumina; a further evidence of the limited mixing is the presence of some traces of corundum ($\alpha$-$Al_2O_3$, from unreacted $\gamma$-$Al_2O_3$). Formulations C and D were produced in order to insure the completion of the reaction between the polymer and the nano-sized filler. The water medium was loaded with nano-composite powders, i.e. with powders (with a dimension of about 1-20 μm) from dried dispersion of alumina nanoparticles in acetone solution of MK polymer. Fig. 3a (C only coating material) shows that in this case the silicone/alumina reaction was complete, as in the case of monoliths. Fig. 4 shows that the coating from formulation C, applied to SiC foams, was homogeneous and thick (Fig. 4a), but still contained some cracks, after thermal treatment (Fig. 4b and Fig. 4c). Mullite powders, in formulation D, were added to minimize the cracks. The addition of mullite micro-powders did not affect the homogeneity of the material (Fig. 4d and Fig. 4e), and it was actually effective in limiting the crack development (Fig. 4f). A further improvement was achieved by a slight increase of the processing temperature, from 1350°C to 1400°C (see different magnifications in Fig. 4g,h,i).

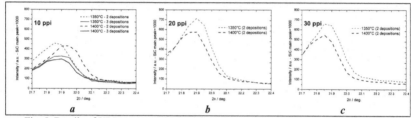

Fig. 5. Details of XRD patterns obtained from foams after oxidation (ground samples).

Fig. 5 shows preliminary results for oxidation tests performed on foams coated with formulation D. The graphs refer to the height of the cristobalite main peak in X-ray diffraction patterns of foams, with different pore density, which were ground after oxidation. This height could be considered as a semi-quantitative indication of the quantity of cristobalite that formed during the oxidation treatment. It may be noted that the height of the cristobalite peak (compared with the height of the SiC main peak, which remains unchanged) decreased with increasing processing temperature, with increasing number of coating cycles and with decreasing pore density. The increase in the processing temperature, from 1350 to 1400°C, and of the number of coating cycles, from two to three, improved the continuity and the thickness of the coating, respectively. The effect of the pore density is probably simply an artifact; in fact, the foams were oxidized in the form of small blocks which were cut from larger coated samples, thus breaking many cell struts. The number of broken struts obviously increased with increasing pore density: this means that foams with high pore density were intrinsically more prone to oxidation through the broken ends of the struts (the SiC foams were prepared by replication, and had hollow struts, which were hardly infiltrated by the precursor material during coating). Foams with small pore density minimized the effect of broken struts, and showed how the mullite layer was effective in protecting the foam against oxidation. The condition which minimized the cristobalite formation - 10 ppi foam, 3 coating cycles, treatment at 1400°C (see Fig. 5a) - was associated to a weight increase of about 2%; while uncoated foams, in the form of large blocks, normally feature a weight increase of 3.5-4%. Although some other oxidation tests (on large samples) are needed for a complete characterization, nano-filled silicones can be effectively considered a promising solution for the preparation of anti-oxidation coatings.

CONCLUSION

The use of a nano-sized alumina filler in conjunction with a silicone allowed the production of phase-pure mullite monoliths and coatings, after heat treatment at 1350°C in air. Coatings were deposited on Si-SiC disks and SiC foams; oxidation tests showed that they were effective in protecting such substrates against oxidation (1300°C for 100 h in static air). The addition of zirconia nano-sized powders led to the formation of dense zirconia-toughened mullite monoliths. Fracture toughness values showed a three-fold increase over pure mullite samples, because the process proposed enabled to homogeneously disperse the nano-sized grains of tetragonal zirconia within the mullite matrix.

REFERENCES

[1] I.A. Aksay, D.M. Dabbs, and M. Sarikaya, Mullite for Structural, Electronic, and Optical Applications, *J. Am. Ceram. Soc.,* 74, 2343-58 (1991).
[2] H. Schneider, J. Schreuer, B. Hildmann, Structure and properties of mullite – A review, *J. Eur. Ceram. Soc.,* 28, 329-44 (2008).
[3] H. Schneider, S. Komarneni, Mullite, John Wiley and Sons (2005) 93-112.
[4] H. Schneider, B. Saruhan, D. Voll, L. Merwin, A. Sebald, Mullite precursors phases, *J. Eur. Ceram. Soc.,* 11, 87-94 (1993)
[5] E. Bernardo, P. Colombo, E. Pippel, J. Woltersdorf, Novel Mullite Synthesis Based on Alumina Nanoparticles and a Preceramic Polymer, *J. Am. Ceram. Soc.,* 89, 1577-83 (2006).
[6] E. Bernardo, P. Colombo, S. Hampshire, SiAlON-based Ceramics from Filled Preceramic Polymers, *J. Am. Ceram. Soc.,* 89, 3839-42 (2006).
[7] E. Bernardo, E. Tomasella, P. Colombo, Development of Multiphase Bioceramics from a Filler-containing Preceramic Polymer, *Ceram. Int.,* (2008) in press.
[8] D. Suttor, H.-J.Kleebe, G. Ziegler, "Formation of mullite from filled siloxanes," *J. Am. Ceram. Soc.,* 80, 2541-48 (1997).
[9] J. Anggono and B. Derby, Intermediate Phases in Mullite Synthesis via Aluminum- and Alumina-Filled Polymethylsiloxane, *J. Am. Ceram. Soc.,* 88, 2085-91 (2005).
[10] F. Griggio, E. Bernardo, P. Colombo, G.L. Messing, Kinetic Studies of Mullite Synthesis from Alumina Nanoparticles and a Preceramic Polymer, *J. Am. Ceram. Soc.,* 91, 2529-33 (2008).
[11] K.N. Lee, Current status of environmental barrier coatings for Si-Based ceramics, *Surface and Coatings Technology,* 133-134, 1-7 (2000).
[12] L. Lutterotti, Maud - Materials Analysis Using Diffraction 20 February 2007. Online. http://www.ing.unitn.it/~maud/
[13] G.R. Anstis, P. Chantikul, B.R. Lawn and D.B. Marshall, A critical evaluation of indentation techniques for measuring fracture toughness: I, direct crack measurement, J. Am. Ceram. Soc., 64, 533-38 (1981).
[14] D.J. Green, An introduction to the mechanical properties of ceramics, Cambridge University Press, Cambridge 1998.
[15] P.F. Becher, K.B. Alexander, A. Bleier, S.B. Waters, W.H. Warwick, Influence of $ZrO_2$ Grain Size and Content on the Transformation Response in the $Al_2O_3$-$ZrO_2$ (12 mol% $CeO_2$) System, *J. Am. Ceram. Soc.,* 79 (1993) 657-63.

# FUNCTIONALLY GRADED CERAMICS DERIVED FROM PRECERAMIC POLYMERS

Martin Steinau
University Erlangen-Nuremberg, Department of Materials Science, Chair of Glass and Ceramics
Erlangen, Germany

Nahum Travitzky
University Erlangen-Nuremberg, Department of Materials Science, Chair of Glass and Ceramics
Erlangen, Germany
University Erlangen-Nuremberg, Institute of Advanced Materials and Processes (Zentralinstitut für neue Materialien und Prozesstechnik ZMP), Fürth, Germany

Timo Zipperle
University Erlangen-Nuremberg, Department of Materials Science, Chair of Glass and Ceramics
Erlangen, Germany

Peter Greil
University Erlangen-Nuremberg, Department of Materials Science, Chair of Glass and Ceramics
Erlangen, Germany
University Erlangen-Nuremberg, Institute of Advanced Materials and Processes (Zentralinstitut für neue Materialien und Prozesstechnik ZMP), Fürth, Germany

ABSTRACT
    Functionally Graded Ceramics (FGC) were fabricated by lamination of preceramic polymer tapes, consisting of polysilsesquioxanes, methyle-triethoxysilane, silicon, and silicon carbide. The tapes were cast using the Doctor-Blade method. After drying at room temperature, the obtained tapes were partially cross-linked at 120 °C. Lamination was carried out by warm pressing at 230°C in an uniaxial press where the individual layers bonded together and formed a bulk material. In order to avoid cracking and warpage, rate controlled pyrolysis of the laminates was applied up to 950°C in flowing $N_2$. After this step the samples were treated at 1200 °C at an $N_2$ pressure of 1 MPa to improve the mechanical properties. Laminates, each consisting of eight layers with graded and ungraded grain-size were produced. Graded structures with a smaller particle-size in the outer layers than in the core showed a 25% higher flexural strength than ungraded structures. The increase of the flexural strength was explained by the generation of compressive surface stresses, due to a different shrinkage and coefficients of thermal expansion (CTE) of the individual layers during heat treatment. An improvement of the fracture toughness $K_{IC}$ could only be observed after the pyrolysis at 950 °C. It could be concluded that, in this case, crack propagation is determined by compressive stresses in the matrix. After the matrix crystallization above 1200 °C, crack propagation was controlled by the deflection of the cracks at the matrix/particle interface. Therefore, laminates with larger grains are favored.

INTRODUCTION
    Polymer-derived ceramics are widely known as bulk or fiber-reinforced engineering materials for several years now. Fabrication techniques were developed from methods used in the polymer industry, such as injection molding, or uniaxial warm pressing[1-6]. The potential of polymer-derived thick-film ceramic substrates for semiconductor applications has successfully been demonstrated by Cromme et al.[7]. In this work laminates with a graded composition were fabricated by casting three layers with different silicon-contents over each other, using the Doctor-Blade technique.

Preceramic polymers like polysilsesquioxanes, polycarbosilanes or polysilazanes show a strong shrinkage when heated to temperatures above 600 °C, as the organic side chains evaporate and a porous, silicon-oxy-carbide glass is formed. Shrinkage and porosity can be reduced by the addition of inorganic filler materials. Depending on their reactivity during the heat treatment, the filler materials can be classified as "passive" or "active". Passive fillers are ceramic materials, such as SiC, $Al_2O_3$, $B_4C$ $Si_3N_4$ $TiO_2$, TiN or $ZrO_2$. During the pyrolysis they behave inert and are embedded in the microstructure without showing reactions with the other constituents. Contrarily, Active filler materials show reactivity with components of the system. For examples Si, Ti, B, Cr, Al, intermetallic compounds and alloys, such as $MoSi_2$, FeSi, $CrSi_2$ or FeSiCr, can react with residual carbon in the structure or $N_2$ during the heat treatment above 1200 °C. By the use of certain combinations of fillers shrinkage, porosity, flexural strength and many other material properties can be tailored as a ceramic micro-composite is formed. This process is called "Active Filler Controlled Pyrolysis" (AFCOP). Polymer derived ceramics, produced with this "Active Filler Controlled Pyrolysis" (AFCOP) showed a flexural strength of up to 400 MPa. However, an amount of 30 to 50 vol.% active filler is needed to achieve the high mechanical strength described in literature[1-3]. Active fillers such as FeSi, $CrSi_2$ or $MoSi_2$ that lead to the best properties, however, are very expensive raw materials and their strong reactivity with carbon can result in warpage of the specimens during the heat treatment. Therefore, the addition of the cheaper passive fillers, e.g. SiC or $Al_2O_3$, is necessary to reduce costs and rejections. The consequence of this are reduced mechanical strength and fracture toughness.

An introduction of compressive surface stresses may increase the properties of PDC. This effect has successfully been applied for other technical ceramics like SiC, $Si_3N_4$ or $Al_2O_3$ by lamination of tapes with different composition[8-10]. It was shown that symmetrical laminates can be toughened by inducing compressive stresses in certain layers by choosing a laminate composition with different coefficients of thermal expansion (CTE) in the different layers. The changes in compressive and tensile stresses depend on the mismatch of CTE's and Young's moduli and on the thickness ratio of layers (even/odd). It was shown, that it is important to generate the compressive stresses not only at or near the surface, but also inside the material, to effectively hinder internal cracks or flaws[11,12]. Another way of increasing fracture toughness is the utilization of phase changes in different layers. It was shown that fracture toughness could be raised from 7 to 30 MPam$^{-1/2}$ in laminates where in the outer layers pure $ZrO_2$ was added to $Al_2O_3$ while in the inner layers $Y_2O_3$-stabilized zirconia was used. The phase transformation of the unstabilized $ZrO_2$ to the monoclinic form in the outer layers during sintering and cooling, while the 3Y-TZP retained the tetragonal form, lead to high residual compressive stresses near the surface due to the constrained volume expansion of the monoclinic $ZrO_2$[9,13,14].

The primary objective of this work was to achieve similar effects in polymer-derived ceramics if tapes with different composition, thickness or content of active fillers are laminated. To achieve this tapes with different SiC grain sizes, used as passive filler were laminated. Because of the different shrinkage, CTE's of the layers, strengthening effects could occur during the warm-pressing and the heat treatment. The induction of compressive surface stresses can significantly improve the mechanical properties of polymer-derived ceramics with a low amount of active filler. Therefore, warpage and failure of the material during the furnace treatment can be avoided and the costs of the raw materials can be kept low. Since techniques can be applied that require relatively low temperatures and pressures the use of polymer-derived ceramics offers a cheap and flexible alternative. Further on, a wide range of filler materials can be used, offering many opportunities in tailoring the materials properties by the layer stacking design.

EXPERIMENTAL

In this work, a polymer mixture of polymethylsiloxane (PMS, Silres MK, Wacker, Germany), polyphenylsiloxane (PPS, Silres H63C, Wacker, Germany) and methyl-triethoxysilane (MTES, ABCR, Germany) were used in combination with the active filler silicon (Si HQ 0 - 10 μm, Elkem, Germany).

The passive filler material was silicon carbide powder (SiC green, F600, $d_{50} = 8.5\mu m$, [S8.5], F800, $d_{50} = 6.5\mu m$, and NF 2/025 NB, $d_{50} = 1.5 \mu m$, bimodal [S1.5]; ESK-SiC, Germany). The terms in brackets are the abbreviation of the powders used in this paper. To catalyze the cross-linking reaction of the polymers aluminum-acetylacetonate (AlAcac, Merck, Germany) and oleic acid (Merck, Germany) were used. The constituents were mixed in a way that the final composition of the slurries was 19 vol.% PMS, 20 vol.% PPS, 20 vol.% MTES, 10 vol.% Si, 0.5 % AlAcac, 0.5% oleic acid and 30 vol.% SiC. The slurries were homogenized for 12 h at 60 rpm in a ball mill using $ZrO_2$-balls with 10 mm diameter. After agitation, the slurries were de-gassed in a rotary evaporator (VV2000 Heidolph, Germany) for 60 min at 0.03 MPa. Rheological behavior was tested at shear-rates between 0.1 $s^{-1}$ to 1000$s^{-1}$ using the cone/disk method in a rotational rheometer (UDS 200 Paar Physika, Germany). Tape-casting was performed using the Doctor-Blade Technique on a tape casting machine (Simatic OP7, Siemens Germany). The blade height was adjusted to 1 mm. Drawing speed was set to 700 mm/min. The tapes were dried at room temperature for 12 h and then partially cross-linked in a cabinet dryer for 2 h at 120 °C. After this step, the tapes had a thickness of 650 $\mu m$ to 750 $\mu m$ due to the evaporation of the solvent. Before lamination the tapes were cut into rectangles with the length of 210 mm and a width of 105 mm. Eight sheets were stacked and pressed in a preheated mold at 230 °C for 20 min and 25 MPa pressure in an uniaxial press (HL60, Lindner, Germany). After the pressing step, temperature of the mold was held at 230 °C for 40 min before demolding and cooling of the plates in a preheated furnace from 230 °C to room temperature in 1 h. The stacking order of the different laminates is given in table I. To differentiate between the compositions, the individual tapes are named after the used passive filler, as the rest of the composition is identical. Below table I a schematic depiction of the laminate's stratification is given. For mechanical testing, bars with the dimensions of 53 x 6 x 6 mm³ were cut out of the laminates using a diamond saw (Brillant 250, ATM, Germany). From the dimensional change after the thermal treatment the volumetric shrinkage was calculated. Linear shrinkage and the coefficient of thermal expansion (CTE) were measured using a dilatometer (DIL 402 ES/3, Netzsch, Germany). To optimize the heating rates for the pyrolysis, thermogravimetric analyses (TGA) were performed in a Thermogravimetric Analyzer 951 (TA Instruments, USA). For conversion of the thermoset green bodies into the ceramic material, a tailored two-step pyrolysis approach was applied, where the process was performed in nitrogen atmosphere including over pressure treatment. In the first stage between 20 and 950 °C, the Si-, O-, C- and H-containing polymer is converted into an inorganic amorphous Si-O-C structure, while the organic side groups of the polymer are evaporated. This step was performed in an electrically heated furnace (GLO 40, Gero GmbH) with a constant mass loss rate, to avoid cracking due to the development of high internal pressures in the samples caused by uncontrolled gas evaporation. In the second stage, the material was additionally heated with 3 K/min to 1200 °C with a dwell time of 3 h at 1MPa $N_2$ in a gas-pressure furnace (FPW 180/250-2-2200-100-PS, FCT Anlagenbau, Germany).

After pyrolysis, density of the materials was measured by the Archimedes method (AG 204, Mettler Toledo, Germany) and a He-pyknometer (Akkupyc 1330 V2.04N, Micromeritics GmbH, Germany). Microstructural and phase analyses were conducted by scanning electron microscopy (SEM, Quanta 200, FEI, Czech. Rep.) and X-ray diffractometry (XRD, Kristalloflex D500, Siemens, Germany), respectively. Mechanical strength was measured in a standard testing machine (Instron 5565, Instron Corp., USA) after DIN 53452 in four-point bending with spans of 20 and 40 mm. The dimenstions of the sample for bending tests were 50 x 5.5 x 5 mm³. The tensile surfaces of the samples were polished to a 6 $\mu m$ diamond finish prior to bending. Fracture toughness was determined according to the Indentation Strength in Bending (ISB) method[15]. A Vickers indentation with 98.1 N load (HV10) was placed on the polished surface of a sample. Afterwards a four-point bending test was performed and the fracture toughness calculated. For the measurements samples with the dimensions 30 x 5.5 x 5 mm³ were used. The spans of the bending test were 10 and 20 mm. The modulus of elasticity was calculated from the longitudinal sound-propagation velocity measurements with these samples by the impulse excitation technique (Buzz-o-Sonic, BuzzMac Software, USA)[16]

Table I. Stacking order of the fabricated laminates

| Lam-1 (S8.5) | Lam-2 (S1.5) | Lam-3 | Lam-4 |
|---|---|---|---|
| S8.5 | S1.5 | S1.5 | S8.5 |
| S8.5 | S1.5 | S1.5 | S8.5 |
| S8.5 | S1.5 | S8.5 | S1.5 |
| S8.5 | S1.5 | S8.5 | S1.5 |
| S8.5 | S1.5 | S8.5 | S1.5 |
| S8.5 | S1.5 | S8.5 | S1.5 |
| S8.5 | S1.5 | S1.5 | S8.5 |
| S8.5 | S1.5 | S1.5 | S8.5 |

RESULTS AND DISCUSSION

Preceramic polymers offer the opportunity of the adhesive free lamination of tapes using warm pressing at temperatures below 250 °C. This could be achieved by the partial cross-linking of the tapes at 120 °C. During this treatment the formation of the duroplastic polysilsesquioxane was not completed and an material with properties between a thermoplastic an duroplastic polymer was formed. The tapes still showed a temperature dependent viscosity and therefore can be laminated easily by the application of pressure at a temperature where the cross-linking reaction is finalized. This resulted in the formation of a duroplastic / ceramic composite material. Figure 1 shows a laminate where tapes with a $d_{50}$ of 8.5 μm were laminated with tapes where an SiC with a $d_{50} = 1.5$ μm was used. It could be observed, that the individual tapes were joined together by the polymer-matrix. Because of the adhesive-free lamination, the material was very resistant against shear forces and delamination.

To obtain a ceramic material the pyrolytic conversion of preceramic polymers was performed at temperatures above 400 °C. At these temperatures the decomposition of the polymer-phase began and a dimensional shrinkage as well as a mass loss, due to the evaporation of the alkyl and phenyl side chains was observable. During this sequence a significant gas-fraction is formed which can cause cracking of the sample, since high pressures can develop in the material. Also warpage of the can occur as inhomogeneities in the material lead to different shrinkage. To reduce these effects a rate-controlled pyrolysis was applied during the thermal treatment up to temperatures of 950 °C. The process is on the basis of the rate controlled binder burn-off for ceramics[17] and other binder-containing materials[18]. From a thermogravimetric analysis, as shown in figure 2a, and a constant assumed mass-loss rate in percent per minute, a heating curve is deviated and applied to the furnace circle, figure 2b:

$$dT/dt = (dm/dt)/(dm/dT) \tag{1}$$

Experiments showed that the number of rejections could be reduced by 30% when this method was applied. After the conversion of the preceramic polymer phase into the amorphous Si-O-C-matrix, the samples were heated in a second step up to 1200°C in 1 MPa $N_2$. In this process the transformation of the Si-O-C-glass into a crystalline ceramic phase, accompanied by the formation of SiC from the active filler Si and the free carbon in the structure began. To achieve a complete conversion of the Si-O-C, Si and C, higher treatment temperatures above 1350 °C are needed.

Figure 3 shows the volume shrinkage of the four laminated after heating to 950 °C and 1200 °C in $N_2$-atmosphere. It could be shown that shrinkage during the organic – inorganic conversion depends on the average particle size of the passive filler. The laminate with the larger SiC-grains shows a increased shrinkage compared to the one with the finer filler. Also, the CTE during cooling is higher if the coarse SiC-filler is used. The linear CTE of Lam-1 is $8\times10^{-6}$ $K^{-1}$ while Lam-2 has a CTE of $6\times10^{-6}$ $K^{-1}$.The two "graded" structures show a shrinkage in between the values of the monolithic samples. The decreased shrinkage of the samples with the smaller grain size can be explained by the different packing density of the fine and the coarse SiC, although in all samples the same volume fraction of filler is used. The higher packing density of fine grains leads to a constriction of the shrinkage of the material. The formation of a different amount and distribution of the porosity can also be an explanation.

Figure 1. Laminate structure of tapes with two different SiC grain-sizes in cross-linked state. The dashed line shows the interface between the tape with an average SiC grain-size of 8.5 μm and one with 1.5 μm.

Figure 4 shows the microstructure of the interface between the coarse and the fine layer of the laminates after the treatments at 950°C and 1200 °C. After the treatment at 950°C in flowing $N_2$ the amorphous Si-O-C phase (dark gray) the crystalline SiC particles (light gray) and the metallic Si-grains were observable, figure 4 a. After the heat treatment of 3 h at 1200 °C in 1 MPa $N_2$ the silicon has already partially reacted to SiC and the matrix is crystallized and densified, figure 4b. By EDX-analysis also traces of the Si-O-N phase could be detected. These phase transformation contributed to the improvement of the flexural strength.

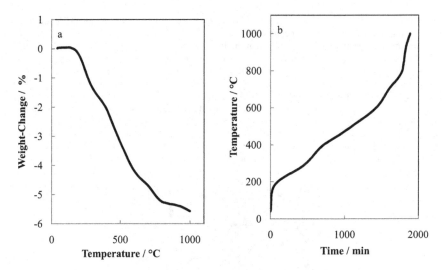

Figure 2. TGA-curve (a) with a heating rate of 3 K/min in $N_2$ and deviated heating curve with a constant mass-loss of 0.003 % / min (b).

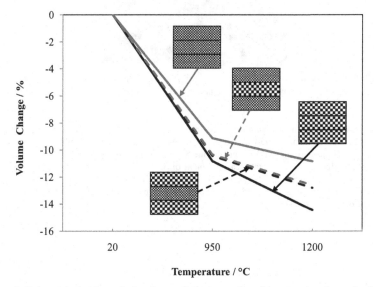

Figure 3. Volumetric shrinkage during the pyrolytic conversion of preceramic polymer laminates between room temperature and 1200 °C in $N_2$ atmosphere.

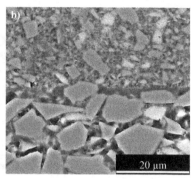

Figure 4. Microstructure of a laminate consisting of tapes with an average SiC diameter of 1.5 μm in the cover layers and 8.5 μm in the core. The sample in figure 4a was pyrolyzed rate-controlled up to 950 °C. The sample in figure b was pyrolyzed for 3 h in 1 MPa $N_2$ at 1200 °C.

The cracking surface at the interface between the coarse and the fine layers of samples thermally treated at 950°C and 1200 °C is shown in figure 5. At the interface almost no deflection of the crack was observable. It can also be seen that the crack propagated mostly in the Si-O-C matrix and at the matrix-SiC interface. Accordingly, the surface of the layer with the coarse SiC is much rougher and the crack is deflected stronger in layers with coarse SiC.

The densification of the microstructure at the higher heat treatment temperature has strong influence on the mechanical properties. The bending strength, $\sigma_c$, of the four laminates is shown in figure 6. After the pyrolysis at 950 °C, the samples had about 15 % open porosity and therefore had a relatively low mechanical strength. A lamination of tapes with the fine SiC in the outer layers lead to an increase of the bending strength of about 25 %. If tapes with the coarse SiC on the outside were used, $\sigma_c$ is decreased by the same amount. Similar results were obtained after the thermal treatment at 1200 °C. Samples treated at 1200 °C showed an increased bending strength, as the amount of total porosity is reduced to ~ 5 %. The glass matrix was densified at its crystallization had already began. The strengthening effect in the laminate with the fine SiC on the outside is probably caused by the different shrinkage and the different CTEs during heating and cooling of the individual layers. As the inner layers with the coarse SiC try to shrink stronger than the outer layers, it can be assumed, that compressive stresses induced to the outer layers contributed to a strengthening of the material.

Regarding the fracture toughness, $K_{IC}$, a strengthening effect by compressive stresses could only be observed after the pyrolysis at 950 °C, figure 7. However, after the heat treatment at 1200 °C, the laminate with the finer layers on the outside showed a lower fracture toughness than the samples with the coarse SiC in the outer layers. This is an indication that the $K_{IC}$ is determined not only by the compressive stresses in the matrix but also by the crack deflection at the matrix – particle interface depending on the bonding between the different phases. It can be assumed, that after the treatment at 950 °C, fracture toughness could be dominated by the compressive stresses, since the strength of the Si-O-C / SiC interface is relatively low. If the material is heated to higher temperatures, the strength of the matrix – particle interface becomes stronger and therefore compressive stresses in the matrix could have less influence on the crack growth.

Figure 5. Fracture surface of a laminate consisting of tapes with an average SiC diameter of 1.5 μm in the cover layers and 8.5 μm in the core. The sample in figure (a) is pyrolyzed rate-controlled up to 950 °C. The sample in figure (b) is heat treated for 3 h in 1 MPa $N_2$ at 1200 °C

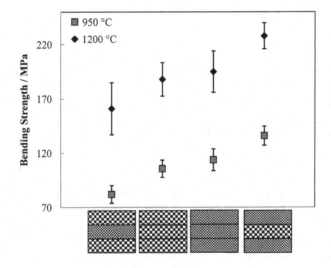

Figure 6. Bending strength of laminates with graded and ungraded average grain size heat treated at 950°C in 0.1 MPa $N_2$ and 1200 °C at 1 MPa $N_2$.

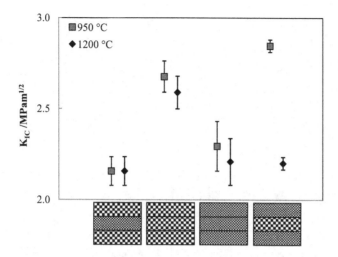

Figure 7. Fracture toughness of laminates with graded and ungraded average grain size heat treated at 950°C in 0.1 MPa $N_2$ and 1200 °C at 1 MPa $N_2$.

To show that crack propagation is determined by the two effects – internal stresses and crack deflection - Vickers indents were placed into the cross section of samples near the interface between the fine and the coarse layers. The cracks emerging from the edges of the indentation were analyzed using SEM. It could be observed, that in the layers with fine SiC, cracks propagating in the direction of the tape-normal, were much shorter than those growing along the tape plane, figures 8a and b. Contrarily, in layers with the coarse SiC, cracks propagating in the direction of the tape normal were longer than those in the direction of the tape plane, figure 8c and d. This observation was made for both temperatures. The crack extension in the different layers also gave an idea of the deflection of the crack at the matrix – particle interface. While in layers with fine SiC the cracks grew almost straight through the material, they were deflected stronger in the layers with the coarse SiC. Thus, a crack propagating in layers with coarse filler particles needed more energy to grow than in layers with fine SiC.

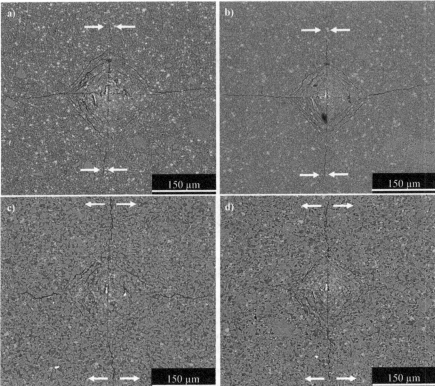

Figure 8. Vickers indentations in the cross section of a laminate with fine SiC filler on the outside and coarse SiC in the inner layers. The samples in figures (a) and (c) were pyrolyzed at 950 °C in 0.1 MPa $N_2$; the samples in (b) and (d) were heat treated at 1200 °C in 1MPa $N_2$. In layers with fine SiC cracks parallel to the tape normal are constrained by compressive stresses, while in the layers with the coarse SiC they can grow longer due to tensile stresses.

CONCLUSIONS

In the present study the potential of functionally graded ceramic laminates derived from preceramic polymers was shown. By lamination of polymer tapes fabricated with the Doctor-Blade technique, adhesive free laminates with bulk properties could be fabricated. It can be supposed that the generation of compressive surface stresses, was realized by the lamination of tapes with different shrinkage and CTE because of a different filler grain-size, while the filler content was held constant. Flexural strength could be improved by 25 % if the difference in the grain size between the outer layers and the core of the material had the relation of ~1:6. Vickers indentation into the cross section of the graded laminates showed indications of the existence of compressive stresses in the outer layers, if tapes with fine SiC were placed on the outside. The fracture toughness in these samples could be increased only after the pyrolysis at 950 °C. After the thermal treatment at higher temperatures, $K_{IC}$ is determined more by the crack deflection at the particle – matrix interface than by compressive stresses.

Due to this, structures where filler materials with larger average grain size were used showed higher fracture toughness than laminates with fine SiC.

The results showed that a graded or stepped lamination of polymer derived ceramic tapes with different compositions can lead to an improvement of the flexural strength. As many tape-compositions and layer-combinations are conceivable, it is possible to tailor the laminates properties by the layer stacking desighn. Because of this, the lamination technique offers a wide range for the production of parts with many possible applications beginning from failure tolerant bearings over applications with improved wear resistance to materials with graded thermal properties.

REFERENCES
[1] P.Greil, Active-Filler-Controlled Pyrolysis of Preceramic Polymers," *J. Am. Ceram. Soc.*, **78**, 835–848 (1995).
[2] P. Greil, Near Net Shape Manufacturing of Polymer Derived Ceramics, *J. Eur. Ceram. Soc.*, **18**, 1905–14 (1998).
[3] P. Greil, Polymer Derived Engineering Ceramics, *Adv. Eng. Mater.*, **2**, 339 (2000)
[4] S. Walter, D. Suttor, T. Erny, B. Hahn & P. Greil, Injection Moulding of Polysiloxane/Filler Mixtures for Oxycarbide Ceramic Composites, *J Eur. Ceram. Soc.*, **16**, 387-393 (1996)
[5] Jens Cordelair, Peter Greil, Electrical conductivity measurements as a microprobe for structure transitions in polysiloxane derived Si-O-C ceramics, *Journal Eur. Cer. Soc.* **20**, 1947-1957 (2000)
[6] M. Steinau N. Travitzky, J. Gegner, J. Hofmann, P. Greil, Polymer-Derived Ceramics for Advanced Bearing Applications, *Adv. Eng. Mat.*, **10**, 1141-1146, (2008)
[7] P. Cromme, M. Scheffler, P.Greil, Ceramic Tapes from Preceramic Polymers, *Adv. Eng. Mat.*, **4**, 873-877 (2002)
[8] A.J. Blattner, R. Lakshminarayanan , D.K. Shetty, Toughening of layered ceramic composites with residual surface compression: effects of layer thickness, *Eng. Frac. Mech.*, **68**, 1-7 (2001)
[9] R. Lakshminarayanan , D.K. Shetty, R. Cuttler, Toughening of Layered Ceramic Composites with Residual Surface Compressions, *J. Am. Ceram. Soc.*, **79**, 79-87 (1996)
[10] R.A. Cutler, A.V. Virkar, The Effect of Binder Thickness and Resiual Stresses on the Fracture Toughness of Cemented Carbides, *J. Mater. Sci.*, **20**, 3557-73, (1985)
[11] N. Orlovskaya, M. Lugovy, V. Subbotin, O. Radchenko, J. Adams, M. Chheda, J. Shih, J. Sankar, S. Yarmolenko, Robust Design and Manufacturing of Ceramic Laminates with Controlled Thermal Residual Stresses for enhanced toughness, *J. Mater. Sci.*, **40**, 5483-5490, 2005
[12] M.P. Rao, A.J. Sanchez-Herencia, G.E. Beltz, R.M. McMeeking, F.F. Lange, Laminar Ceramics That Exhibit A Threshold Strength, *Science*, **286**, 102 (1999)
[13] A.V. Virkar, J.L. Huang, R.A. Cutler, Strenghtening of Oxide Ceramics by Transformation Induced Stresses, *J. Am. Ceram. Soc.*, **70**, 164-70, 1987
[14] R.A. Cutler, J.D. Bright, A.V. Virkar, D.K. Shetty, Strength Improvement in Transformation Toughened Alumina by Selective Phase Transformation, *J. Am. Ceram. Soc.*, **70**, 714-18 (1987)
[15] G. Anstis, P. Chantikul, B. Lawn and D.B. Marshall, A Critical Evaluation of Indentation Techniques for Measuring Fracture Toughness: I, Direct Crack Measurements, *J. Am. Ceram. Soc.*, **64**, 533–38 (1981).
[16] M. Radovic, E. Lara-Curzio, L. Rieser, Comparison of Different Experimental Techniques for Determination of Elastic Properties of Solids, *Mater. Sci. Eng.*, **368**, 56-70 (2004)
[17] J.W. Yun, S.J. Lombardo, D.S. Krueger, P. Scheuer, Effect of Decomposition Kinetics and Failure Criteria on Binder Removal Cycles From Three-Dimensional Porous Green Bodies, *J. Am. Ceram. Soc.*, **89**, 176-183 (2006)
[18] B. Derfuss, M. Gruhl, C.A. Rottmair, A. Volek, R.F. Singer, Net-shape production of graphite parts via powder injection molding of carbon mesophase, *J. Mat. Proc. Tech.*, **208**, 444–449 (2008)

GENERATION OF CERAMIC LAYERS ON TRANSITION METALS VIA REACTION WITH SiCN-PRECURSORS

C. Delpero, W. Krenkel, G. Motz
Ceramic Materials Engineering, University of Bayreuth
Ludwig-Thoma-Strasse 36 b
Bayreuth, Germany, 95447

ABSTRACT
        Transition metal carbide and nitride layers have been generated by reaction of transition metals with a SiCN-precursor. Sheets of Hf, Nb and Mo were dip-coated with the polycarbosilazane ABSE and subsequently pyrolysed at 1000°C-1300°C. Carbon and nitrogen from the precursors reacted with the metals and formed carbides and nitrides in the interface area between the ceramic layer and the metal substrate. Transition metal nitride and carbide phases have excellent chemical and physical properties that make them suitable for high temperature applications. Due to the diffusion processes gradient layers of carbides and nitrides are formed which should provide for a good adhesion of the layers. The layers were found to be 2-3µm in thickness after annealing at 1000°C and up to 20µm thickness at 1300°C. Phases of $Nb_2N$, HfN, $HfN_{0.4}$ and $Mo_2C$ were identified via X-ray diffraction. The precursor route is an excellent method to synthesise ceramic coatings on transition metal substrates at comparatively low temperature without complex equipment.

INTRODUCTION
        Transition metal carbides and nitrides are promising materials for use in high temperature applications. Their high hardness, high melting points (see Tab. I), wear resistance and chemical inertness also make them suitable for many other technical applications. Niobium nitride is of special interest because aside of its excellent thermal and chemical properties it also has excellent electrical properties which makes it a possible candidate for superconducting junctions[1]. Hafnium nitride has an extremely high melting point of 3387°C[2] which, combined with its good oxidation resistance in a propulsion environment, makes it interesting as material for hypersonic jet engines as well as for furnace elements or plasma arc electrodes[3]. Molybdenum carbide too has a high melting point, good chemical stability and additionally has excellent properties as a catalyst.

Table I. Properties of NbN, HfN and $Mo_2C$[2]

|  | NbN | HfN | $Mo_2C$ |
|---|---|---|---|
| Melting point | 2400°C | 3387°C | 2520°C |
| Vickers Hardness | 13.3GPa | 16.3GPa | 15.5GPa |

        For the synthesis of these transition metal nitrides and carbides very many different methods exist such as reactive hot pressing[3], cathodic arc physical vapour deposition[4], plasma spraying[5], unbalanced magnetron sputtering[6], pulsed laser ablation[7], radical beam assisted deposition[8] or high-temperature electrochemical synthesis[9]. But these established techniques all require either very high temperatures, high pressures or expensive equipment. Also, the adhesion of the layers on the substrates or their stability under cyclic thermal impact is often limited when deposited with these methods. Another drawback is the fact that it is difficult to coat complex shaped samples with most of these methods.
        Reactive precursor-based coating is an approach for the synthesis for nitrides and carbides that has many advantages over the well established synthesis methods. As shown by Ryu and Raj[10] the temperature required for the formation of nitrides and carbides via the precursor route is much lower

than that required for e.g. reactive hot pressing, which is about $2100°$[3]. It can be realised with simple dip- or spray-coating techniques that require almost no apparative effort and it is also possible to process complex shaped components. In this work it is reported on coatings derived from the reaction of Hf, Nb and Mo with a tailored soluble polycarbosilazane (ABSE) during a thermal treatment at temperatures > 800°C. As already demonstrated for titanium the formation of TiN[10] or gradient layers[11] is expected by reaction of the metals with reactive precursor species.

EXPERIMENTAL PROCEDURE

The ABSE polycarbosilzane is produced in our own institute. It is synthesized by ammonolysis of bis(dichloromethysilyl)ethane in toluene as described by Motz[12] and Traßl[13]. ABSE (Fig. 1) is a colourless, brittle and meltable solid and can be solved in unpolar solvents. The sheets of hafnium, molybdenum and niobium were provided by HMW Hauner, Roettenbach, Germany.

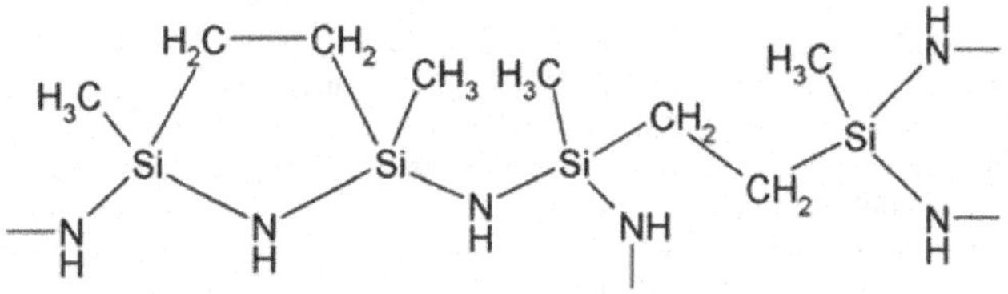

Figure 1. Basic structure units of the ABSE precursor[14]

Samples were coated by dip-coating the metal sheets with a hoisting apparatus in solutions of the ABSE precursor in toluene. Thickness of the precursor layer can be controlled via the hoisting speed, the concentration of the polymer in the solution and the viscosity of the solvent[15]. The samples were handled under a nitrogen atmosphere in a glove box to ensure that samples are free of impurities. Coated substrates were annealed in special atmospheres - nitrogen or argon - for 1h to 10h at 1000°C and 1300°C in a tube furnace (Heraeus RO 10/100).

Cross-sections of the resulting samples were examined with scanning electron microscopy (SEM, Jeol JSM 6400) and energy dispersive X-ray spectroscopy (EDX, Noran System Six Model 300, Thermo Fisher). The composition of the precursor and gradient layers was determined with glow discharge spectrometry (GDOES / Spectruma 750P). Crystalline phases in the gradient layers were identified with X-ray diffraction (Seifert XRD 3000P). Additionally the microhardness of the layers was characterized with a Fisherscope® H100 (Helmut Fischer GmbH & Co.KG Germany).

RESULTS AND DISCUSSION

Nitridation in pure $N_2$ atmosphere

Due to the expected reactivity of the used metals with nitrogen at elevated temperatures, at first the nitridation behaviour of the different metals was examined when heated in a pure nitrogen atmosphere. In Figure 2 it can clearly be seen that both hafnium and niobium show a distinct affinity to forming nitrides at elevated temperatures, while molybdenum shows only very little mass change. At temperatures of 1400°C and above even loss of mass can be observed in the molybdenum sample, which is probably due to MoN sublimating, since MoN has a melting point of only 1750°C[2]. The other two transition metals show a great mass increase at temperatures above 1300°C which proves that these elements form nitride phases. At a temperature of 1000°C though the mass increase is only small which means that the nitridation at this temperature is limited. This is why the scope of this work was to generate nitride phases on these metals at temperatures of 1000°C via the precursor route.

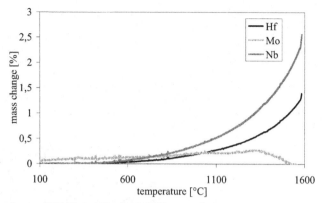

Figure 2. Mass change of Hf, Nb and Mo sheets in $N_2$ atmosphere

Gradient coatings on niobium

Niobium has a melting point of 2468°C which makes it suitable for high temperature applications. Niobium is able to react with every element in the SiCN-precursor forming NbN, NbC or $Nb_2Si$. The formation enthalpy of NbN is far higher than that of NbC or $Nb_2Si$ (Tab. II). Thus the formation of nitride phases after coating Nb with ABSE and pyrolysing the sample is favored. XRD measurements showed that in fact only two nitride phases have formed during pyrolysis (Fig. 3).

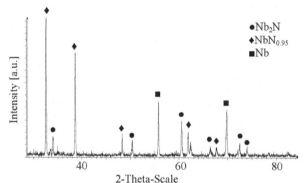

Figure 3. XRD-spectrum of ABSE coated Nb, annealed at 1000°C for 1h in $N_2$

Table II. Formation enthalpies of Nb compounds[2]

|  | NbN | NbC | $Nb_2Si$ |
|---|---|---|---|
| Formation enthalpy | -326.0kJ/mol | -140.6kJ/mol | -97.5kJ/mol |

When the nitrogen diffuses into the metal it can be expected that a hard ceramic gradient layer of NbN forms with the N concentration decreasing with the distance from the surface. This will minimize the thermal mismatch as the thermal expansion gradients will not change drastically from the nitride to the metal. Thus the layers should have a good adhesion to the substrates. To analyze the influence of the atmosphere, ABSE-coated samples were pyrolysed both in nitrogen and argon. That way it can be tested whether the pyrolysis atmosphere affects the formation of the nitride layer.

With XRD measurements $Nb_2N$ and $NbN_{0.95}$ phases could be identified in the samples annealed in nitrogen as well as in argon for one hour at 1000°C, while no phases of NbC or NbSi could be detected in either sample. The chemical composition of the gradient layer was investigated using glow discharge optical emission spectroscopy (GDOES). These measurements confirm the diffusion of nitrogen into the niobium causing the formation of niobium nitrides (Fig. 4). Also, if we compare the nitrogen diffusion of an uncoated Nb sheet in nitrogen annealed with the same parameters it can be seen that the formation of NbN is less pronounced.

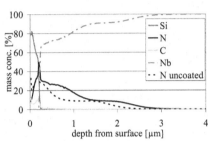

Figure 4. GDOES measurement of ABSE coated Nb, annealed at 1000°C for 1h in $N_2$

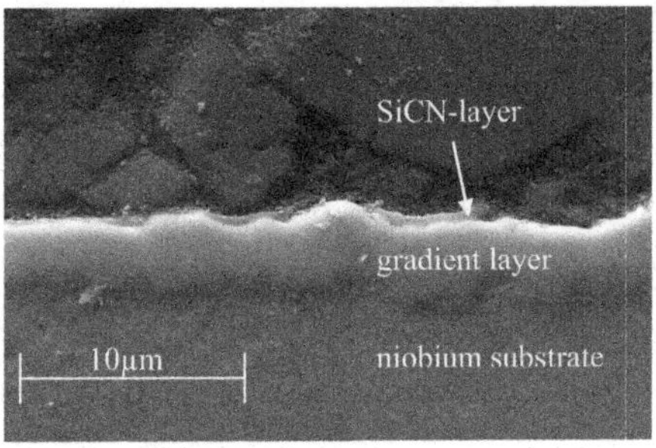

Figure 5. Cross-section of ABSE coated Nb and intermediate gradient layer

The effect of the SiCN layer on the formation of a gradient nitride layer can be seen even better when the sample is pyrolysed at 1300°C. The nitridation of the uncoated part is much less than the nitridation in the interface area between the ceramic layer and the metal. The excellent bonding of the ABSE layer on the substrate is the cause of the remarkable nitrogen diffusion into the niobium. This effect was also observed by Günthner et al. who coated titanium sheets with ABSE to produce gradient layers of Ti nitrides and carbides[11]. A chemically polished REM cross-section also shows the diffusion zone between the SiCN-layer and the metal substrate (Fig. 5). Since the Nb nitride is harder and more chemically stable than the metal, it can be clearly seen in the cross-section as it is etched less than the substrate.

Gradient coatings on hafnium

Hafnium sheets were also coated with the ABSE precursor and pyrolysed at the same conditions as the Nb samples. XRD measurements showed the presence of a HfN and a $HfN_{0.4}$ phase after annealing the ABSE coated Hf sheets for 10h at 1000°C in $N_2$. This is in a good accordance to the expected results as the formation enthalpy of HfN is much higher compared to the formation of HfC or HfSi (see Tab. III).

Table III. Formation enthalpies of Hf compounds[2]

|  | HfN | HfC | $HfSi_2$ |
|---|---|---|---|
| Formation enthalpy | -369.4kJ/mol | -209.6kJ/mol | -226.1kJ/mol |

Characterization via GDOES shows the diffusion of N from the SiCN-layer into the Hf substrate, which relates to the results obtained from X-ray diffraction. The GDOES profile also provides for diffusion of small amounts of carbon into the sample. After nitridation of an uncoated Hf sample in pure $N_2$ atmosphere the nitrogen amount is strongly reduced in comparison to the coated sample (Fig. 6). This result is comparable to formation of nitrides in the Nb samples. To investigate the hardness of the surface, micro-hardness measurements were performed (Fig. 7). It can be seen that compared to an uncoated Hf sheet the hardness on the surface of the coated sample is higher due to the formed ceramic layer. The hardness value only slowly decreases to the level of the bulk material which correlates with the GDOES measurement showing the formation of hard nitride phases.

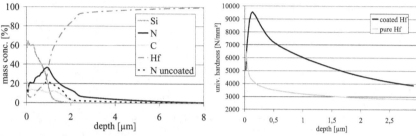

Figure 6. GDOES measurement of ABSE coated Hf, annealed at 1000°C for 1h in $N_2$

Figure 7. Microhardness profile of ABSE coated Hf, annealed at 1000°C for 1h in $N_2$

Gradient coatings on molybdenum

In the preliminary investigations it was found that molybdenum does not tend to form nitrides when heated up to 1500°C in a pure $N_2$ atmosphere. After coating with ABSE and pyrolysing for 10h at 1000°C in $N_2$ an $Mo_2C$ carbide was detected with X-ray diffraction in the molybdenum sample (Fig. 8). This can not be explained with the formation enthalpies because $Mo_2C$ has the lowest formation energy (Tab. IV). But it is known that the free carbon content of ABSE after pyrolysis at 1000°C is about 31%[16], so it can be assumed that this free carbon is responsible for the formation of the carbide phase. This result is in accordance with the investigations of Ryu and Raj[10]. They warm pressed a mixture of 10 vol% molybdenum powder and an SiCN-precursor and annealed the sample for 10h at 1200°C. The resulting metal-ceramic composite only yielded molybdenum carbide phases ($\alpha$-$Mo_2C$ and $\gamma$-$MoC$) in the XRD measurement.

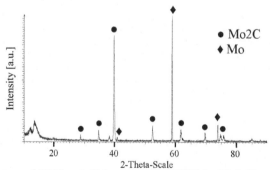

Figure 8. XRD-spectrum of ABSE coated Mo, annealed at 1000°C for 10h in $N_2$

Table IV. Formation enthalpies of Mo compounds[2]

|  | $Mo_2N$ | $Mo_2C$ | $MoSi_2$ |
|---|---|---|---|
| Formation enthalpy | -70kJ/mol | -46kJ/mol | -109.3kJ/mol |

GDOES measurements (Fig. 9) also support the XRD results and demonstrate the carbon diffusion into the Mo substrate. No nitrogen is found in the molybdenum substrate so we can assume that there is only a carbide phase. The composition of about 6.5 weight-% carbon and 93.5 weight-% molybdenum determined with GDOES measurements corresponds very well with the composition of a $Mo_2C$ phase which has 5.9 weight-% carbon. In Figure 10 a cross-section of this sample examined with SEM shows the interface area between the ceramic layer and the metal substrate. Examination with energy-dispersive X-ray spectroscopy also proves the existence of a carbon containing phase between the ceramic layer and the metal which correlates with both the XRD and GDOES measurements.

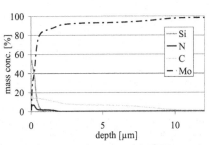
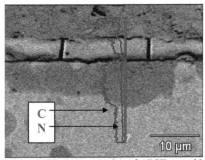

Figure 9. GDOES measurement of ABSE coated Mo, annealed at 1000°C for 1h in $N_2$

Figure 10. SEM cross-section of ABSE coated Mo and intermediate gradient layer

CONCLUSION

Sheets of the transition metals Hf, Mo and Nb were coated with the SiCN-precursor ABSE and annealed for 10h at 1000°C and 1300°C in nitrogen and argon atmospheres. The precursor layer transforms into an amorphous SiCN ceramic and carbon or nitride diffuse into the metal forming gradient carbide and nitride layers. XRD, GDOES and SEM/XRD measurements were used to characterize the samples. By using Hf and Nb sheets exclusively nitride phases were formed in the surface, while for Mo only a carbide phase was detected. The nitride phases were about 2-3μm in thickness while the carbide phase found in Mo is in the range of 6-7μm. Micro-hardness measurements were additionally performed to prove the existence of a gradient layer in the samples.

The precursor technology is a promising route for the synthesis of ceramic layers on transition metals at comparatively low temperatures and without the need of complex equipment. Further investigations will include the characterization of the chemical and mechanical properties of the layers to assess their potential for practical applications as well as to clarify the layer formation.

REFERENCES
[1] V.N. Zhitomirsky, I.Grimberg, L. Rapoport, N.A. Travitzky, R.L. Boxman, S. Goldsmith, A. Raihel, I. Lapsker, B.Z. Weiss, "Structure and mechanical properties of vacuum arc-deposited NbN coatings", *Thin Solid Films*, **326**, 134-142 (1998)

[2] H.O. Pierson, "Handbook of Refractory Carbides and Nitrides: Properties, Characteristics, Processing and Applications", Noyes Publications, Westwood (1996)

[3] M.M. Opeka, I.G. Talmy, E.J. Wuchina, J.A. Zaykovski, S.J. Causey, "Mechanical, Thermal, and Oxidation Properties of Refractory Hafnium and Zirconium Compounds", *Journal of the European Ceramic Society*, **19**, 2405-2414

[4] N. Cansever, "Properties of niobium nitride coatings deposited by cathodic arc physical vapour deposition", *Thin Solid Films*, **515**, 3670-3675 (2007)

[5] Y. Chen, T. Laha, B. Kantesh, A. Agarwal, "Nanomechanical properties of hafnium nitride coating", *Scripta Materialia*, **58**, 1121-1124 (2008)

[6] X.T. Zeng, "TiN/NbN superlattice hard coatings deposited by unbalanced magnetron sputtering", *Surface and Coatings Technology*, **113**, 75-59 (1999)

[7] G. Cappuccio, U. Gambardella, A. Morone, S. Orlando, G.P. Parisi, "Pulsed laser ablation of NbN/MgO/NbN multilayers", *Applied Surface Science*, **109/110**, 399-402 (1997)

[8] N. Hayashie, I.H. Murzin, I. Sakamoto, M. Ohkubo, "Single-crystal niobium nitride thin films prepared with radical beam assisted deposition", *Thin Solid Films*, **259**, 146-149 (1995)

[9] V.V. Malyshev, A.I. Hab, "Electrochemical and corrosion behaviour of titanium with molybdenum-carbide coatings in solutions of sulphuric acid", *Materials Science*, **39**, No. 6, 901-904 (2003)

[10] H.-Y. Ryu, R. Raj, "Selection of TiN as the Interconnect Material for Measuring the Electrical Conductivity of Polymer-Derived SiCN at High Temperatures", *Journal of the American Ceramic Society*, **90**, No. 1, 295-297 (2007)

[11] M. Günthner, Y. Albrecht, G. Motz, "Polymeric and ceramic-like coatings on the basis of SiN(C) precursors for protection of metals against corrosion and oxidation", *Ceramic engineering and science proceedings*, **27**, No. 3, 277-284 (2007)

[12] G.Motz, J. Hacker, G. Ziegler, "Special Modified Silazanes for Coatings, Fibers and CMCs", *Ceramic Engineering & Science Proceeding*, **21**, 307-314 (2000)

[13] S. Trassl, D. Suttor, G. Motz, E. Rößler, G. Ziegler, "Structural Characterisation of Silicon Carbonitride Ceramics derived from Polymeric Precursor", *Journal of the European Ceramic Society*, **20**, 215-222 (2000)

[14] G. Motz, G. Ziegler, "Simple Processibility of Precursor-derived SiCN Coatings by optimized Precursors", *Proceedings of the Seventh Conference & Exhibition of the European Ceramic Society*, **1**, 475-478 (2001)

[15] C.J. Brinker, G.W. Scherer, *Sol-Gel Science*, Academic Press, Inc., Boston (1990)

[16] S. Trassl, G. Motz, E. Rössler, G. Ziegler, "Characterization of the free-carbon phase in precursor-derived SiCN ceramics", *Journal of Non-Crystalline Solids*, **293-295**, 261-267 (2001)

ACKNOWLEDGEMENTS
      This work is financed by the DFG research training group 1229 "Stable and Metastable Multiphase Systems at High Application Temperatures".

# FACILE CERAMIC MICRO-STRUCTURE GENERATION USING ELECTROHYDRODYNAMIC PROCESSING AND PYROLYSIS

Z. Ahmad, M. Nangrejo, U. Farook, E.Stride, M. Edirisinghe
Department of Mechanical Engineering, University College London, Torrington Place, London WC1E 7JE, UK

E. Bernardo, P. Colombo[&,$,*]
University of Padova, Dipartimento di Ingegneria Meccanica – Settore Materiali
via Marzolo, 9, 35131 Padova, Italy

ABSTRACT
   Electrohydrodynamic processing of a pre-ceramic polymer solution (polysiloxane) was used to fabricate various micro-components. With a single needle experimental set-up, sub-micrometer fibers and millimeter-size porous capsules were produced, while using two co-axial needles further enhanced the shaping versatility of the process, leading to the formation of micro-tubes and sub-micrometer hollow capsules. The polymeric samples were cross-linked and then pyrolyzed at 1200°C in $N_2$ producing ceramic micro-parts that retained the morphological features present in the components in the polymeric state.

INTRODUCTION
   The electrohydrodynamic (EHD) process is a versatile fabrication method that has been used to produce very different structures, which may have a wide range of applications in a variety of remits, ranging from biomedical devices to electronic uses. Electrospinning and electrospraying are the most commonly used EHD processing methods and are linked by the elasto-viscosity properties of processing solutions. Electrohydrodynamic-based technologies have encountered a rapid and significant growth, thanks to the ample benefits they offer both in terms of processability and also with regards to the miniaturized scaled structures which can be fabricated. The fact that several factors can be manipulated to influence the process affords a great versatility to the technique. EHD has been recently proposed in conjunction with pre-ceramic polymers for the fabrication of nano-fibers[1-5]. Our previous work demonstrated the possibility to extend the range of morphologies and components that can be produced using this technique, both in the as processed, polymeric state and as ceramic components, after pyrolysis[6-9]. These methods have yielded a range of structures using single needle and co-axially aligned needles during processing. While single needle processing can generate a series of morphologies, co-axially aligned needles can offer a broader range of structures with compartmentalization. For encapsulation of materials, there are certain criteria (e.g. immiscibility of materials under co-flow, alignment of needles) which need to be fulfilled, and this is also the case for successful EHD processing in both single needle and co-axial needles (e.g. conductivity and surface tension).[7]
   In this paper, we briefly discuss the use of several possible approaches for developing polymeric and ceramic miniaturized components with complex morphology using EHD processing.

EXPERIMENTAL
   The preceramic polymer used in this work was a commercially available methyl-silsesquioxane (PMSQ) (MK, Wacker-Chemie, Burghausen, Germany). PMSQ was dissolved in pure ethanol (VWR

---

[&] Department of Materials Science and Engineering, The Pennsylvania State University, University Park, PA 16802
[$] Department of Mechanical Engineering, University College London, Torrington Place, London WC1E 7JE, UK
[*] Member, American Ceramic Society. Corresponding author: paolo.colombo@unipd.it

International Ltd., Poole, UK) with mechanical stirring (900 seconds) to prepare two solutions with different polymer concentrations (18 and 60 wt%). The solutions were subjected to EHD flow or co-flow using sets of stainless steel needles assembled to give the processing device. The needles are coaxially aligned and assembled in a concentric manner (see Figure 1) and they can be utilized simultaneously or alone, depending on the structure that needs to be formed. The smallest needle diameter used in these experiments was 150 μm. The device was coupled to a high voltage supply (Glassman Europe Limited, Bramley, UK). Different materials can be introduced into these individual needles, and in these experiments we used glycerol (Sigma Aldrich, Poole, UK), PMSQ polymer solutions and air. The location of the materials was decided based on the desired product and the number of functional needles, which in this work was restricted to a maximum of two, albeit in other experiments up to three needles were used.[7] Selected solutions were introduced into the needles at a controlled flow rate using syringe pumps (Harvard Apparatus Ltd., Edenbridge, UK). Flow rates varied in the range $1.5 \times 10^{-9}$ to $5 \times 10^{-9}$ $m^3 s^{-1}$. The operating applied voltage ranged between 1-10 kV.

Three substrates were used to collect the polymeric structures produced upon processing; 1) a thin film of liquid (distilled water) on a microscope slide; 2) plain microscope slides and 3) a vial of distilled water. The deposition distance, from the needle tip to the collection surface, when present, was varied from 10 to 20 mm. The collection time was varied between 1 min to > 20 mins, and samples were allowed to dry depending on the desired morphology.

Before pyrolysis, the polymeric structures were crosslinked by immersing them into an aqueous solution of ammonia (7 mol/l; pH 13) for 24 h at 40°C (which is below the glass transition temperature of the preceramic polymer). After drying, the samples were pyrolyzed at 1200°C in $N_2$ (heating/cooling rate 2°C/min; dwelling time 1 h). The microstructural features were evaluated from fracture surfaces using a scanning electron microscope (SEM, Stereoscan 250, Cambridge Instruments, Cambridge, UK).

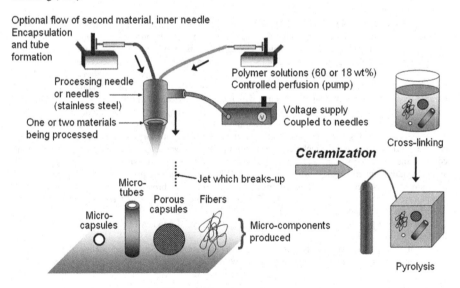

Figure 1. Experimental set-up.

RESULTS AND DISCUSSION

The main aim of this work was to demonstrate the great ease in shaping pre-ceramic polymeric micro-structures using electrohydrodynamic forming, and their subsequent conversion into an identical ceramic format. Therefore, several samples were prepared using either single or co-axial processing using either the 60 or 18 wt% PMSQ solutions. The individual and co-axial needle parameters are given in terms of their outer diameter (OD) and inner diameter (ID).

Firstly, it has to be noted that several parameters can be adjusted to influence the shape (fibers, tubes, bubbles, pots, capsules), dimension (from sub-micrometer to millimeter-size), porosity (amount and type) and surface characteristics (roughness and porosity) of the produced components, and these have been summarized in Table I. In particular, the needle parameters have an impact on the EHD process and influence the dimension of the final product (which can have dimensions < 1 μm). The needle design and parameters can influence the method; and this can have an impact on the stability and cone-jet domain during processing [10,11] The use of multiple needles permits encapsulation of structures and it has been shown that this can increase the morphological variations.[7,12]

| Parameter | Effect |
|---|---|
| Solution concentration | Morphology (fiber/pot/capsule/tube/bubble) |
| Applied voltage | Size distribution, jetting mode and morphology |
| Flow rate | Size distribution and encapsulation ability |
| Collection distance | Drying time, size and jet break-up |
| Collecting medium | Phase mixing, pore generation, surface variation |
| Collecting time | Density and number of samples |
| Solution Phase/Solvent | Immiscibility and encapsulation |
| Number of needles | Number of compartments in structure |

Table I. Processing parameters affecting the characteristics of components produced by EHD.

EHD processing of polymeric solutions, including pre-ceramic precursors, using a single-needle device can result in various morphologies such as droplets, beads or threads, depending on the experimental conditions. In particular, the solution concentration and the polymers molecular weight, plays a pivotal role in such morphological variations under EHD flow. These parameters influence the processing solutions viscosity and the resultant changes have been documented in previous studies.[1-5]

The fundamental change in the visco-elastic properties leads to a transition between electrospraying and electrospinning, which arise owing to the properties of the polymeric solution. During generic EHD processing a solution is drawn from the processing nozzle in the form of a cone which then either breaks up (for a solution with low viscosity) to give droplets, or remains intact and overcomes continuous bending and whipping motions (for a solution with high viscosity).[13] Using a single needle (ID, OD: 1900 and 2800 μm) and the 60 wt% PMSQ-ethanol solution, fibers with diameters ranging from 1 to 25 μm are formed (Figure 2a). In this instance, electrospinning of the 60 wt% solution was achieved at a flow rate of 200 μl/min and an applied voltage of 5.5 to 6.9 kV. The fibers produced were dense, because only one needle was used and there was no significant solvent effect such as rapid evaporation, which is known to produce hollow fibers, for instance when using sol-gel based solutions[14]. Alternatively, the same pre-ceramic polymer solution can be used to prepare highly porous spheroidal capsules, as shown in Figure 2b (1 to >3 mm in diameter, as shown in insert, Figure 2b). This process requires the application of a low voltage (< 5 kV) and flow rates between 90 and 300 μl/min, with a processing needle diameter of 330 μm. The pore size of these capsules can be controlled by the applied voltage, which has to be maintained below 5 kV, after which there is a transition into fiber formation. The porous capsules were collected in a vial of water which permits, together with the chemical and physical characteristics of the preceramic polymer (such as insolubility in water, hydrophobicity and molecular weight), the formation of extensive porosity within the components.

When co-flows are used under EHD flow, the morphologies can be made to vary drastically resulting in particle encapsulation and microtube formation (Figure 2c). For example, using the same 60 wt% PMSQ solution and co-flowing this with glycerol under an electric field can result in the generation of microtubes. This was achieved by perfusing glycerol into the inner needle (ID, OD: 1500 and 900 μm) at a flow rate of 100 μl/min and PMSQ solution in the outer needle (ID, OD: 1900 and 2800 μm) at 200 μl/min. The process required an applied voltage of 6.6-7.3 kV and the inner diameter of these tubes ranged from 2 to 5 μm.

Reducing the concentration of the polymeric solution to 18 wt% and co-flowing this under an electric field with air can result in the preparation of fluid filled microcapsules (Figure 2d). In this instance, the 18 wt% PMSQ solution was perfused in the outer needle at 300μl/min (OD, ID:1100 and 685 μm) and the air flow was also fixed at the same value (OD, ID: 300 and 150 μm). Under these flow conditions the capsules were generated at an applied voltage of 5.7 kV and the samples were collected in a vial of water. After drying, their mean size was determined to be 6 μm. The surface of these capsules is clearly different to those of the larger capsules prepared using a single needle.

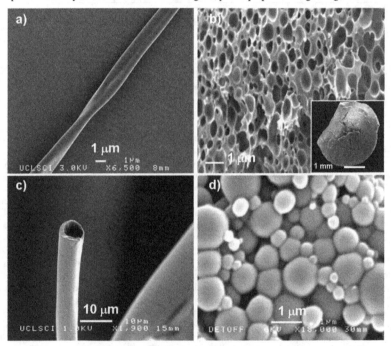

Figure 2. Micro-components obtainable using the device in Fig. 1. a) single needle – dense fibers; b) single needle - porous capsules; c) co-axial needles - micro-tubes d) co-axial needles - fluid filled capsules. Samples were in the as-produced (polymeric) un-pyrolyzed state.

The structures obtained using a polymethylsilsesquioxane preceramic polymer could be used as fabricated, in the polymeric state, for instance in drug-delivery or biomedical applications. For components to be used in miniaturized systems (e.g. MEMS, micro heat exchangers, filters) that have to withstand harsh environments or high temperatures, they can be ceramized by pyrolysis. In order for them to retain the shape during pyrolysis, crosslinking of the preceramic polymers has to be achieved, by post-forming treatments such as heating at a suitable temperature (a catalyst has to be added to the polymer before shaping) or by immersing the component in a highly alkaline water-based solution, as in the present experiments. To elucidate this, two structures formed from the earlier processing experiments (micro-tubes and porous capsules) were cross-linked and then pyrolyzed. In Figure 3 SEM images are reported, showing how fine details can be retained, after pyrolys, in the final ceramic components.

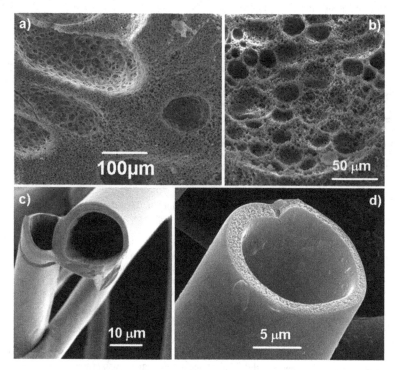

Figure 3. Retention of the morphology after pyrolysis in ceramic micro-components produced by EHD processing. Porous capsules: a) polymeric, as prepared (after curing); b) ceramic, after pyrolysis (showing the retention of multi-scale porosity). Micro-tubes: a) polymeric, as prepared (after curing); b) ceramic, after pyrolysis (showing the presence of porosity in the tube wall).

CONCLUSIONS

Electrohydrodynamic processing, using either a single-needle or a co-axial two-needle experimental set-up, enabled the fabrication of samples widely varying in shape, such as fibers, porous capsules, tubes or bubbles. The size of the samples varied from sub-micron to a few millimeters. Cross-linking and pyrolysis led to the formation of ceramic micro-components retaining the morphological features present in the polymeric state.

ACKNOWLEDGMENTS

The authors acknowledge the EPSRC platform grant EP/E045839, Royal Academy of Engineering and the Leverhulme Trust (F/07 134/BL) for supporting this work. Collaboration between UCL and the University of Padova was funded by the Royal Society and we wish to acknowledge the exchange visits made possible by this grant. PC acknowledges the support of the European Community's Sixth Framework Program through a Marie-Curie Research Training Network (PolyCerNet MRTN-CT-019601).

REFERENCES

[1] D. T. Welna, J. D. Bender, X. Wei, L. G. Sneddon and H. R. Allcock, "Preparation of Boron-Carbide/Carbon Nanofibers from a Poly(norbornenyldecaborane) Single-Source Precursor via Electrostatic Spinning," *Adv. Mater.*, **17** 859–62 (2005).

[2] W.M. Sigmund, V. Maneeratana and P. Colombo, "Ceramic nanofibers for liquid and gas filtration and other high temperature applications", US Patent Application 60/894338, March 12, 2007.

[3] D.-G. Shin, D.-H. Riu and H.-E. Kim, "Web-type silicon carbide fibers prepared by electrospinning of polycarbosilanes," *J. Ceram. Proc. Res.*, **9** 209–14 (2008).

[4] S. Sarkar, A. Chunder, W. Fei, L. An and L. Zhai, "Superhydrophobic Mats of Polymer-Derived Ceramic Fibers," *J. Am. Ceram. Soc.*, **91** 2751–55 (2008).

[5] B. M. Eick and J. P. Youngblood, "SiC nanofibers by pyrolysis of electrospun preceramic polymers," *J. Mater. Sci.*, **44** 160–65 (2009).

[6] U. Farook, M.J. Edirisinghe, E. Stride and P. Colombo, "Novel co-axial electrohydrodynamic in-situ preparation of liquid-filled polymer shell microspheres for biomedical applications," *J. Microencap.*, **25** 241–47 (2008).

[7] Z. Ahmad, H.B. Zhang, U. Farook, M. Edirisinghe, E. Stride, P. Colombo, "Generation of multi-layered structural features for biomedical applications using a novel three-needle co-axial device and electrohydrodynamic flow," *J.Roy. Soc. Interf.*, **5** 1255–61 (2008).

[8] M. Nangrejo, Z. Ahmad, E. Stride, M.J. Edirisinghe and P. Colombo,, "Preparation of Polymeric and Ceramic Capsules by a Novel Electrohydrodynamic Process," *Pharm. Dev. Technol.*, **13** 425–32 (2008).

[9] M. Nangrejo, U. Farook, Z. Ahmad, E. Bernardo, P. Colombo, E. Stride, M. Edirisinghe, "Electrohydrodynamic Forming of Porous Ceramic Capsules from a Preceramic Polymer," *Mater. Lett.*, **63** 483–85 (2009).

[10] C.H. Chen, M.H.J. Emond, E.M. Kelder, B. Meester, J. Schoonman, Electrostatic sol-spray deposition of nanostructured ceramic thin films. *J. Aerosol. Sci.* **30**(7), 959 (1999).

[11] L.Li & A.Tok. Electrospraying of water in the cone-jet mode in air at atmospheric pressure. *Int. J. Mass Spectrom.* **272**, 199 (2008).

[12] I. G. Loscertales, A. Barrero,I. Guerrero, R. Cortijo, M. Marquez, A. M. Gañán-Calvo. Micro/Nano Encapsulation via Electrified Coaxial Liquid Jets. *Science* **295**, 1695 (2002).

[13] Q.P.Pham, U. Sharma, A.G. Mikos. Electrospinning of polymeric nanofibers for tissue engineering applications: a review. *Tissue Eng.* **12** 1197, (2006).

[14] V. Maneeratana, W.M. Sigmund, "Continuous hollow alumina gel fibers by direct electrospinning of an alkoxide-based precursor," *Chem. Eng. J.*, **137** 137–143 (2008).

# DEVELOPMENT OF Si-N BASED HYDROGEN SEPARATION MEMBRANE

Keita Miyajima[1,4], Tomokazu Eda[1], Haruka Ohta[1], Yasunori Ando[1], Shigeo Nagaya[2], Tomoyuki Ohba[3], and Yuji Iwamoto[4]

[1]Research and Development Center, Noritake Co., Limited, 300 Higashiyama, Miyoshi-cho, Aichi, Japan.

[2]Electric Power Research and Development Center, Chubu Electric Power Co., Inc. 20-1 Kitasekiyama, Ohdaka-cho, Nagoya, Aichi, Japan.

[3]Minamata Research Center, Chisso Corp., 1-1 Noguchi-cho, Minamata, Kumamoto, Japan.

[4]Department of Frontier Materials, Graduate School of Engineering, Nagoya Institute of Technology, Gokiso-cho, Showa-ku, Nagoya, Aichi, Japan.

## ABSTRACT

Polymer/ceramic conversion behaviors under $NH_3$ atmosphere were investigated for the polysilazanes having different molecular weight (Mn) of 1300, 2800, 3000 and 4800 (g/mol). The mass spectra detected for $CH_4$, $NH_3$ and $H_2$ gas evolutions from the polysilazanes revealed that the decomposition of $CH_3$ groups followed by Si-N bond formations proceeded up to 650 °C, while subsequent densification of the amorphous Si-N occurred at 700 to 1100 °C. The results of $N_2$ sorption analysis for the 650 °C heat-treated samples revealed that the highest specific surface area was achieved by use of the polysilazane with the highest Mn (4800 g/mol), and the this polymer-derived material having a mean pore diameter below 2 nm showed a high thermal stability up to 950 °C. This polymer-derived amorphous Si-N membrane synthesized on an asymmetric $Si_3N_4$ porous support exhibited a single $H_2$ gas permeance, $3.6 \times 10^{-8}$ (mol $Pa^{-1}m^{-2}s^{-1}$) with the $H_2/N_2$ selectivity of 479 at 200 °C. The high-temperature $H_2$-permselectivety at 600 °C evaluated using a mixed gas of $H_2/N_2=1$ resulted in the $H_2$ gas permeance, $5.0 \times 10^{-8}$ (mol $Pa^{-1}m^{-2}s^{-1}$) with the $H_2/N_2$ selectivity of 170.

## INTRODUCTION

Hydrogen based energy systems or devices are potential candidates to solve the growing environmental problems such as $CO_2$ emission. Therefore, hydrogen demand will rapidly increase in near future. Today, about 50 % of hydrogen has been produced through steam reforming of natural gas. In this step, methane ($CH_4$, main constituent of natural gas) is catalytically reacted to form $H_2$:

$$CH_4 + H_2O \iff 3H_2 + CO - 206kJ/mol$$

This reaction is endothermic, and the $CH_4/H_2$ conversion efficiency is limited by thermodynamic equilibrium. Therefore, the reaction is normally carried out at high temperatures(>700 °C). It is well known that the equilibrium can be shifted to the product side, if $H_2$ can be removed selectively from the steam reforming reaction by a $H_2$ separation membrane. As a result, higher conversion of $CH_4$ to $H_2$ can be achieved even at significant lower temperature (approximately <600

$^{o}$C) [1-3]. To develop such a highly efficient reactor, high $H_2$ perm-selectivity and durability at high temperature are required for hydrogen separation membranes. Therefore, micro-porous ceramic membranes are the most promising candidates. Ceramic hydrogen separation membranes based on amorphous $SiO_2$[4-9], $SiC$[10-13] and silica-alumina composite[14,15] have been developed by many researchers. As a dominant mechanism for the $H_2$ permeation thorough amorphous $SiO_2$ membranes, it is proposed that $H_2$ molecules adsorb in the solubility sites and then jump randomly from the site to site through passageways between the sites. The passageways are thought to be formed by combinations of 5-, 6-, 7- or 8-membaered rings often existed within Si-O linkages[16,17]. We considered that it is essential to develop novel $H_2$-permselecitive amorphous silicon nitride (Si-N) based membranes by controlling such a precursors-derived lower-dense amorphous network formation. In this work, the polymer/ceramic conversion behaviors under $NH_3$ atmosphere were investigated for polysilazanes, then, gas permeation properties were studied on the polysilazane-derived Si-N based membrane fabricated on a porous $Si_3N_4$ support. The effect of molecular weight of polysilazanes on the microporous amorphous Si-N-Si network formation is also discussed from a view point to develop novel Si-N based membranes for high-temperature separation of $H_2$.

EXPERIMENTAL

Polysilazanes (Chisso corp., Japan) having different number-average molecular weight (Mn) =1300, 2800, 3000 and 4800 (g/mol) were used. The thermal behaviors of the polysilazanes were investigated under $NH_3$ atmosphere at 550 to 1500 $^{o}$C. The gas evolutions during the heat treatment was monitored by TG-DTA/GC-MS system (Rigaku corp., Japan) using He as a carrier gas. The heat-treated polysilazanes were characterized by the specific surface area and pore diameters determined by the $N_2$ sorption analysis (Shimazu corp., Japan). All the samples were heat-treated at 650 $^{o}$C under $NH_3$ atmosphere, followed by an additional heat treatment at 950, 1100 and 1300 $^{o}$C under $N_2$ atmosphere. For the synthesis of Si-N based hydrogen separation membranes, a porous $Si_3N_4$ tubular support (length, inner and outer diameter of 120 mm, 12 mm and 10 mm) consisting of porous $Si_3N_4$ support layer (mean pore diameter 1000 nm, porosity 35%) and a porous $Si_3N_4$ intermediate layer (mean pore diameter 100 nm, porosity 50 %) was fabricated at our laboratory[18]. The hydrogen separation layer was synthesized by dip-coating of a toluen solution of the polysilazane with Mn=4800 g/mol. The coating process was repeated until the desired thickness was achieved, and converted into the amorphous Si-N layer by heat treatment at 650 $^{o}$C for 1 hr under $NH_3$ atmosphere.

A schematic diagram of the experimental set up to measure $H_2$ perm-selectivity is shown in Fig. 1. In order to evaluate the permeance of He, $H_2$, $CO_2$ and $CH_4$, each single gas was fed from the outside of a cylindrical membrane module at 0.2 MPa, and maintained at 200 $^{o}$C by the electric furnace. The premeance of each gas was calculated from the flow rate of the permeated gas, the relative pressure between the feed side and the permeate side. In order to evaluate the effect of temperature on the $H_2$ permeance and the selectivity, a mixtured gases ($H_2$/$N_2$=1 (molar ratio)) was fed from the outside of the cylindrical membrane module at 0.4 MPa, and the temperature was controlled at 200 to 600 °C. The flow rate of feed gas was controlled at 3 L/min. The flow rate and composition of the permeated gas were measured using a flow meter and the gas chromatography, respectively, and then permeance of $H_2$ and $N_2$ were calculated.

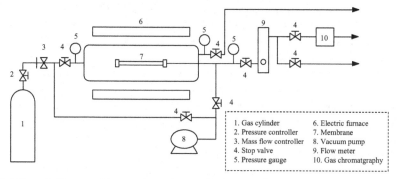

1. Gas cylinder      6. Electric furnace
2. Pressure controller     7. Membrane
3. Mass flow controller   8. Vacuum pump
4. Stop valve          9. Flow meter
5. Pressure gauge      10. Gas chromatgraphy

Fig. 1. Schematic diagram of the experimental set up for the evaluation of gas permeation properties.

RESULT AND DISCUSSION

As a typical result, the mass spectra detected for gas evolutions from the polysilazane (Mn=1300g/mol) during the heat treatment under $NH_3$ atmosphere were shown in Fig. 2. The evolution of $CH_4$ and $NH_3$ was observed at 600 to 850 °C, while that of $H_2$ was found in a slightly higher temperature range from 700 to 1100 °C.

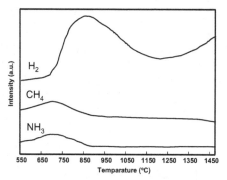

Fig 2. Mass spectra detected during heat treatment under $NH_3$ atmosphere of polysilazane (Mn=1300 g/mol).

Figure 3 shows the effect of the heat treatment temperature on the (a) specific surface area and (b) mean pore diameter of the polysilazane-derived amorphous Si-N. These results indicate that the condensation of polysilazane, and the subsequent densification of the polymer-derived amorphous Si-N-Si network proceeded above 650 °C. Except for the sample derived from the polysilazane with the highest Mn=4800 g/mol, an apparent pore growth begun to started at 950 °C, and the resulting mean pore diameters were approximately 3 to 5 nm. The chemical structure of polysilazanes used in this work was $[(Me_2Si)_2NH]_x[MeSiHNH]_y[MeSiN]_z$[19]. By heating to 650 °C, almost all the methyl groups (Me-Si-) was decomposed with the evolution of $CH_4$, which leading to the formation of $NH_2$-Si-, and subsequent partial condensation to yield -Si-NH-Si-. At 650 °C, the -Si-N-Si- network was thought to be loose, because the formation of the -Si-N-Si- linkages was not fully completed, and there still could be a certain amount of non-bridging nitrogen atoms. The $H_2$ evolution in the range of 700~1100 °C indicated further condensation reactions to form dense amorphous -Si-N-Si- network (Fig. 4).

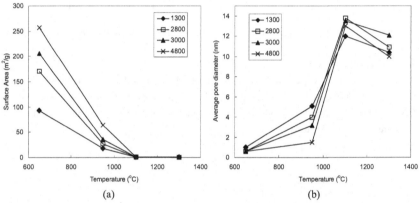

(a)            (b)

Fig. 3. Effect of temperature range on the surface area and average pore diameter of polysilazane having different number-average molecular weights.

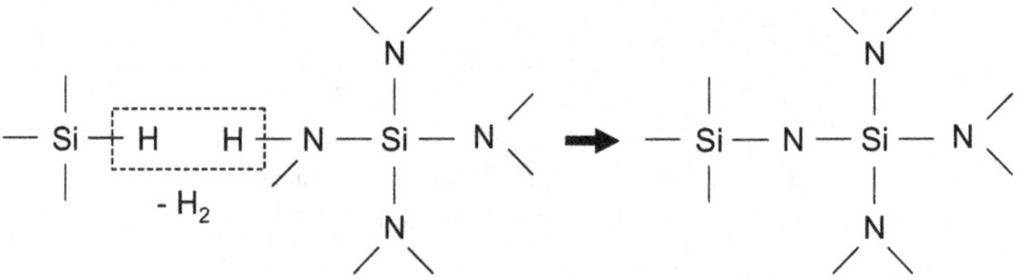

Fig. 4. Condensation reaction to form amorphous Si-N-Si network[19].

As shown in Fig. 3, the specific surface area of the 650 °C-heat treated sample was apparently increased consistently with the molecular weight of the starting polysilazane. Moreover, the microporous structure of the sample derived from the polysilazane with the highest Mn (4800 g/mol) was found to be thermally stable up to 950 °C. The mechanism has not been clarified in detail, however it is apparent that the molecular weight of the polysilazane is an important key parameter to control the condensation in the polymer state and subsequent densification of amorphous Si-N. From these results, it is confirmed that the polysilazane having Mn: 4800 g/mol and heat treatment at 650 °C under $NH_3$ atmosphere are suitable for synthesizing the hydrogen separation layer.

Figure 5 shows the cross-sectional SEM image of the Si-N based hydrogen separation layer successfully synthesized on the porous $Si_3N_4$ support. Crack-free thin layer with a thickness of approximately 200 nm was observed on the outer surface of the porous support.

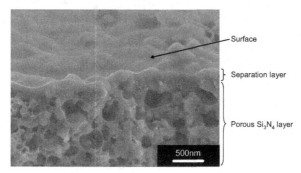

Fig. 5. Cross-sectional SEM image of Si-N based hydrogen separation membrane.

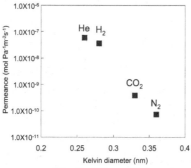

Fig. 6. Single gas permeances of He, $H_2$, $CO_2$ and $N_2$ through amorphous Si-N membrane derived from polysilazane (Mn=4800 g/mol).

The gas permeation properties of the polysilzane-derived Si-N membrane were shown in Figs. 6 and 7. The single gas permeances at 200 °C of He, $H_2$, $CO_2$ and $N_2$ were measured to be $5.9 \times 10^{-8}$, $3.6 \times 10^{-8}$, $3.8 \times 10^{-10}$ and $7.4 \times 10^{-11}$ (mol $Pa^{-1}m^{-2}s^{-1}$), respectively (Fig. 6). The permeances of the smaller gas molecules below 0.3 nm in size (He and $H_2$) were much higher than those of other gas molecules with larger dimensions, and the $H_2/CO_2$ and $H_2/N_2$ selectivities were 94 and 479, respectively. The temperature dependence of the gas permeances evaluated by using a mixed gas ($H_2/N_2$ =1 (molar ratio)) revealed that the $H_2$ permeance increased with increasing temperature, and reached $5 \times 10^{-8}$ (mol $Pa^{-1}m^{-2}s^{-1}$) at 600 °C, while the $N_2$ permeance slightly decreased to be $2.9 \times 10^{-10}$ (mol $Pa^{-1}m^{-2}s^{-1}$), and the $H_2/N_2$ selectivity at 600 °C was determined as 170 (Fig. 7).

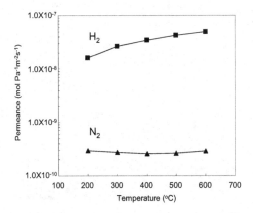

Fig. 7. Temperature dependence of $H_2$ and $N_2$ permeances through the Si-N membraneevaluated using a mixed gas of $H_2/N_2$=1 (molar ratio).

Both He and $H_2$ could permeate through the membrane composed of the polysilizane-derived amorphous Si-N-Si network, and their permeations could be activated at higher temperatures. On the other hand, the permeations of other larger gas molecules ($CO_2$ and $N_2$) were essentially restricted. The observed $CO_2$ and $N_2$ permeations could be due to the existence of a very small amount of defects having several nanometers, and governed by Knudsen diffusion or viscous flow.

The polysilzane-derived amorphous Si-N membrane synthesized in this study was found to show a relatively high $H_2$-permselectivity. Moreover, this polymer-derived membrane material exhibited an excellent thermal stability, and thus can be expected as a new candidate material for developing highly efficient ceramic membranes for high-temperature separation of $H_2$.

CONCLUSION

In this study, the polymer/ceramic conversion behaviors under $NH_3$ atmosphere were investigated for polysilazanes $\{[(Me_2Si)_2NH]_x[MeSiHNH]_y[MeSiN]_z\}$ having different Mn of 1300, 2800, 3000 and 4800 (g/mol). Then, a novel $H_2$-permselective amorphous Si-N membrane was synthesized on a tubular asymmetric $Si_3N_4$ porous support. The results are summarized as follows:

(1) The mass spectra detected for the gas evolutions from the polysilazanes during the heat treatment showed that, up to 650 °C, the decomposition of $CH_3$ groups and subsequent Si-N bond formations occurred. Further densification of the amorphous Si-N-Si network proceeded at 700 to 1100 °C.

(2) $N_2$ sorption analysis for the 650 °C heat-treated samples revealed that the molecular weight of the starting polysilazane was found as one of important key parameters to control the condensation in the polymer state and subsequent densification of the polymer-derived amorphous Si-N. As a result, the highest specific surface area was achieved for the material derived from the polysilazane having the highest Mn of 4800 g/mol. Moreover, this polymer-derived material exhibited the smallest mean pore diameter below 2 nm, and an excellent thermal stability up to 950 °C.

(3) The membrane derived from the polysilazane having the largest Mn achieved a relatively high single $H_2$ gas permeance, $3.6 \times 10^{-8}$ (mol $Pa^{-1}m^{-2}s^{-1}$) with the $H_2/N_2$ selectivity of 479 at 200 °C. The high-temperature $H_2$-permselectivety at 600 °C evaluated using a mixed gas of $H_2/N_2=1$ resulted in the $H_2$ gas permeance, $5.0 \times 10^{-8}$ (mol $Pa^{-1}m^{-2}s^{-1}$) with the $H_2/N_2$ selectivity of 170. These results indicated that the polysilazane-derived amorphous Si-N membrane could be one of candidates for the development of a novel highly efficient membrane reactor for $H_2$ production.

REFERENCES

[1]Jarosch and H. I. de Lasa, Novel Riser Simulator for Methane Reforming Using High Temperature Membranes, *Chem. Eng. Sci.*, **54**, 1455-1460 (1999).

[2]E. Kikuchi, Y. Nemoto, M. Kajikawa, S. Uemiya and T. Kojima, Membrane Reactor Application to Hydrogen Production, *Catal. Today*, **56**, 97-101 (2000).

[3]N. Ito and K. Haraya, A Carbon Membrane Reactor, Catal Today, **56**, 103-111 (2000).

[4]T. Okubo and H. Inoue, Introduction of Specific Gas Selectivity to Porous Glass Membranes by Treatment with Tetraethoxysilane, *J. Membr. Sci.*, **42**, 109-117 (1989).

[5]A. K. Prabhu and S. T. Oyama, Hydrogen Selective Membranes: Application to the Transformation of Greenhouse Gases, *J. Membr. Sci.*, **176**, 233-248 (2000).

[6]D. Lee, S. T. Oyama, Gas Permeation Characteristics of a Hydrogen Selective Supported Silica Membrane, *J. Membr. Sci.*, **210**, 291-306 (2002).

[7]B. N. Nair, T. Okubo and S. Nakao, Structure and Separation Properties of Silica Membrane, *Membrane*, **25**, 73-85 (2000).

[8]S. Gopalokrishnan, Y. Yoshino, M. Nomura, B. N. Nair and S. Nakao, A hybrid Processing Method for High Performance Hydrogen-Selective Membranes, *J. Membr. Sci.*, **297**, 5-9 (2007).

[9]Y. Iwamoto, Precursor-Derived Ceramic Membranes for High-Temperature Separation of Hydrogen, J. Ceram. Soc. Japan, 115, 947-954 (2007).

[10]R. J. Ciora, B. Fayyaz, K. T. Liu, V. Suwanmethanond, R. Mallada, M. Sahimi and T. T. Tsotsis, Preparation and Reactive Applications of Nanoporous Silicon Carbide Membranes, *Chem. Eng. Sci.*, 59, 4957-4965 (2004).

[11]H. Suda, H. Yamauchi, Y. Uchimaru, I. Fujiwara and K. Haraya, Preparation and Gas Permeation Properties of Silicon Carbide-Based Inorganic Membranes for Hydrogen Separation, *Desalination*, 193, 252-255 (2006).

[12]B. Elyassi, M. Sahimi and T. T. Tsotsis, Silicon Carbide Membranes for Gas Separation Applications, *J. Membr. Sci.*, 288, 290-297 (2007).

[13]F. Chen, R. Mourhatch, T. T. Tsotsis and M. Sahimi, Experimental Studies and Computer Simulation of the Preparation of Nanoporous Silicon-Carbide Membranes by Chemical-Vapor Infiltration/Chemical-Vapor Deposition Techniques, *Chem. Eng. Sci.*, 63, 1460-1470 (2008).

[14]N. Nishiyama, M. Yamaguchi, T. Katayama, Y. Hirota, M. Miyamoto, Y. Egashira, K. Ueyama, K. Nakanishi, T. Ohta, A. Mizusawa and T. Satoh, Hydrogen-Permeable Membranes Composed of Zeolite Nano-Blocks, *J. Membr. Sci.*, 306, 349-354 (2007).

[15]Y. Gu, P. Hacarlioglu and S. T. Oyama, Hydrothermally Stable Silica-Alumina Composite Membranes for Hydrogen Separation, *J. Membr. Sci.*, 310, 28-37 (2008).

[16]S. T. Oyama, D. Lee, P. Hacarlioglu and R. F. Saraf, Theory of Hydrogen Permeability in Nonporous Silica Membrane, *J. Membr. Sci.*, 244, 45-53 (2004).

[17]P. Hacarlioglu, D. Lee, G. V. Gibbs and S. T. Oyama, Activation Energies for Permeation He and $H_2$ through Silica Membranes: An ab Initio Calculation Study, *J. Membr. Sci.*, 313, 277-283 (2008).

[18]K. Tsunoda, Y. Ando, S. Nagaya and H. Seo, JP Patent 2005-270716 (2005).

[19]E. Kroke, Y-L. Li, C. Konetschhny, E. Lecomte, C. Fasel and R. Riedel, Silazane derived ceramics and related materials, *Mater. Sci. Eng.*, R26, 97-199 (2000).

# POROUS POLYMER DERIVED CERAMICS DECORATED WITH IN-SITU GROWN NANOWIRES

Cekdar Vakifahmetoglu, Paolo Colombo[#,*,&]
Dipartimento di Ingegneria Meccanica - Settore Materiali, Università di Padova, via Marzolo, 9, 35131 Padova, Italy

ABSTRACT

Cellular SiOC and SiCN ceramic foams containing both open and closed porosity were fabricated from preceramic polymers by varying parameters such as the type of the precursor, pyrolysis temperature and atmosphere. Nanowires were obtained, directly upon heating, on the surface of the macro-porous components by catalyst-assisted reactions occurring during pyrolysis. Cobalt chloride was used as catalyst source. An influence of the pyrolysis temperature and atmosphere on the development of the 1D nanostructures was found, in dependence also of the type of precursor used.

## 1. INTRODUCTION

A porous component containing pores of two or more length scales is referred as a material with hierarchical porosity. Different types of porosity may exist according to the range of pore sizes that are involved in the porous structure, i.e. bimodal size distribution (micro-meso, meso-macro, micro-macro), or trimodal (micro-meso-macro). Graded or oriented porosity is also sometimes required for specialized applications. Such components are of significant technological interest, and are used in several industrial processes and household products. Applications include catalysis, filtration (of liquids or gases), extraction, separation, sorption and scaffolds for biological applications.[1] In general, it can be said that a macroporous ceramic framework offers chemical and mechanical stability, as well as high convective heat transfer, high turbulence, low pressure drop and a high external mass transfer rate due to interconnections between the macropores.[2] The solid walls surrounding these pores can be modified to provide the functionality for a given application (such as chemical affinity towards specific pollutants, high specific surface area, surface roughness, etc.). In the past two decades, the synthesis and functionalization of one-dimensional nano-structural materials (nanotubes, nanowires, etc.) have received steadily growing interests due to their unique and often superior properties compared to that of their bulk and/or microscale counterparts.[3] So far, a great number of manufacturing techniques have been developed for the production of these materials.[3] Of these techniques, the use of preceramic polymers seems to be a promising one due to the great tailorability of their chemical composition on a molecular scale and ease of processing. It has been shown that various types of nanostructures with different chemical composition could be produced from preceramic polymers without the use of any transition metal additives as catalyst.[4-12] Although recently great progress has been made in the use of preceramic polymers via Catalyst-Assisted-Pyrolysis (CAP) to both improve the final yield and to obtain special types of nanostructures,[13-21] relatively few studies

---

[#] and Department of Materials Science and Engineering, The Pennsylvania State University, University Park, Pennsylvania 16802
[*] Member, American Ceramic Society.
[&] Author to whom correspondence should be addressed. e-mail: paolo.colombo@unipd.it

focused on producing porous components decorated with 1D nanostructures,[13; 16; 22; 23] In this paper, a production route for macrocellular ceramic monoliths decorated with 1D nanostructures, obtained by the catalyst assisted pyrolysis of ordinary commercial preceramic polymers, is proposed.

## 2. EXPERIMENTAL PROCEDURE

Porous ceramics were produced by using commercially available preceramic polymers. Poly(methyl)silsesquioxane (MK, Wacker Chemie AG, Burghausen, Germany; denoted as PMS) powder (96wt%) was mixed by ball milling with azodicarbonamide (ADA, 1 wt%), serving as a physical blowing agent, and a transition metal chloride ($CoCl_2$, 3 wt%), acting as the catalyst source. The mixed batch was then transferred to an oven and the temperature was increased to 90°C for 1h, and then curing was obtained at 250°C (2°C/min heating rate). Dwelling at this temperature for 5h enabled to form a porous infusible polymer (ADA decomposes around 210°C[24]). In order to obtain SiCN based materials, a liquid poly(methyl-vinyl)silazane (Ceraset™ VL20, KiON Corporation, Clariant, USA) was mixed with the same chemicals (blowing agent and catalyst source) employed for the polysiloxane; the samples were prepared under Ar atmosphere, using standard *Schlenk* techniques. All the chemicals were added to the precursor under magnetic stirring at room temperature. The homogenous mixture, after 24h of stirring, was poured in aluminum tray and cured at 300°C for 4h (with 2°C/min heating and cooling rate) under $N_2$ atmosphere (99.999% pure). All porous polymer monoliths were then separately pyrolyzed under both $N_2$ and Ar atmosphere (both 99.999% pure) by heating in an alumina tube furnace (2°C/min heating rate) to 1250 or 1400°C; samples were kept at the intended temperature for 2h, after which cooling was performed at the same heating rate.

Thermal analysis (TG-DTA) measurements were carried out up to 1500 °C both under Ar and $N_2$ (Netzsch STA 429, Selb, Germany; 2°C/min heating rate) on the already cured samples. The true density was measured from finely ground ceramic powder using a He-Pycnometer (Pycnomatic ATC, Porotec). Open and closed porosity of the sintered ceramics were determined by the Archimedes principle (ASTM C373-72), using xylene as buoyant medium. The morphological features of the samples were analyzed from fresh fracture surfaces using a scanning electron microscope (JSM-6300F SEM, JEOL, Tokyo, Japan). SEM images were subsequently analyzed with the ImageTool software (UTHSCSA, University of Texas, USA) to quantify the cell size and cell size distribution. The raw data obtained by image analysis were converted to 3D values to obtain the effective cell dimension through the well-known stereological equation: $D_{sphere}= D_{circle}/0.785$.[25]

## 3. RESULTS AND DISCUSSION

PMS precursors can be cured thermally at temperatures >100°C in air[11; 26; 27]. The mineralization of these precursors takes place predominantly between 400 and 800°C[28] and a Si-O-C material forms which remains amorphous at least up to pyrolysis temperatures of 1400°C.[29] Increases in the pyrolysis temperature results in phase separation/decomposition/crystallization of the metastable matrix.[30] TGA analysis measurements, displayed in Fig. 1a&b, are in good agreement with the literature data. The final ceramic yield was only slightly affected by the incorporation of the cobalt catalyst and values of ~90% were obtained whatever the used atmosphere (Ar or N2) was. There was also no clear difference in DTA plots when Co is added or not and whatever the working atmosphere was (the specimens were already cross-linked before analysis).

XRD spectroscopy was performed in order to identify the phase evolution, and the results for the samples pyrolyzed both under $N_2$ and Ar atmosphere at 1250 and 1400°C are shown in Fig. 2. The samples obtained at 1250°C displayed a characteristic diffraction pattern of amorphous $SiCxOy$, with a broad hump in the 20-30 degrees range (2θ) and the peak at 26.50° (2θ) which could be related to free carbon (labeled C in the graphs).[30] Experiments performed at 1000°C showed that cubic Co (JCPDS #15-0826) was the primary cobalt phase (data not shown for the brevity). When increasing the thermolysis temperature, reaction with the silicon-based matrix occurred resulting in a cobalt silicide phase mixtures which was observed by XRD. While thermolysis at 1250°C yielded $Co_2Si$ (JCPDS #04-0847) as the main silicide phase, at 1400°C under both pyrolysis atmosphere CoSi (JCPDS #50-1337) was observed (predominant silicide phase for $N_2$ thermolysis), indicating that the silicidation reaction occurs through the phase sequence: cubic Co → $Co_2Si$ → CoSi, in agreement with previous studies.[31; 32] The formation of cobalt carbide or cobalt silicate compound was not observed by XRD analysis.

Increasing the thermolysis temperature decreased considerably the intensity of amorphous silica and carbon related peaks, and improved crystallization, similarly to what reported for a Ni-containing polysiloxane.[33] While the sample treated under Ar at 1400°C shows only broad features for β-SiC (JCPDS # 29-1129), the one treated under $N_2$ contained well defined crystalline features of $Si_2N_2O$ (JCPDS # 47-1627), together with small amount of β-SiC.

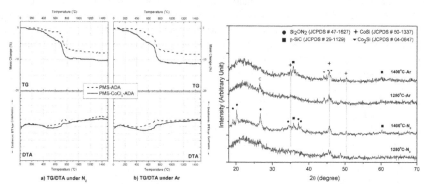

Figure 1. TGA/DTA results for PMS samples pyrolyzed under a) $N_2$, and b) Ar.

Figure 2. XRD plots for PMS samples pyrolyzed under $N_2$ or Ar, at 1250 and 1400°C.

Formation of a $Si_2N_2O$ (silicon oxynitride) phase has been shown to occur after pyrolysis of a mixture of the same polysiloxane and silicon (Si) under $N_2$.[11; 34] The authors showed that the reaction of SiO gas (released from the precursor during pyrolysis) and $N_2$ (the heating atmosphere) formed $Si_2N_2O$, and that a further increase in the temperature resulted in a continuous increase in the absorption of nitrogen up to the melting point of Si ($Tm_{(Si)}$~ 1414°C). It should be noticed that in these studies only a small amount of $Si_2N_2O$ was detected after pyrolysis in $N_2$ at temperatures lower than the melting temperature of the silicon powder additive. In the present study, we have demonstrated that the addition of Co catalyst allows the formation of a well defined $Si_2N_2O$ crystalline phase without the need of any other reactive filler, such as Si. Porous monoliths pyrolized at 1400°C under Ar were comprised of an amorphous SiOC matrix containing micro/nano-crystalline β-SiC, whereas the samples heat treated under $N_2$ were comprised of amorphous SiOC containing micro-crystalline β-SiC and a rather well defined crystalline silicon oxynitride phase (in the nanowires, as evidenced by TEM investigations, not reported here).

The formation of a large amount of porosity in the samples, after curing, was due to the decomposition of the blowing agent. The presence of some closed porosity (see later) could be attributed to the limited amount of volatiles/oligomers released during the PMS curing (in contrast to other polysiloxane precursors, such as a poly-methyl- phenyl-siloxane,[35] and to a poor match between the decomposition temperature of ADA and the viscosity or the preceramic polymer at this temperature.[26] It should be noted, however, that this temperature can be modified by the addition of suitable activators.[36] The morphology of the ceramic foam made by PMS-CoCl$_2$-ADA closely resembled, in macroscale, that of the polymeric thermosets not containing the catalyst (data not shown), namely showing both open and closed porosity with dense struts, similarly to previous investigations.[26] However, the change in pyrolysis atmosphere caused differences in the final products at the nanoscale. Upon 1250°C heating under $N_2$, entanglements of nanowires were created, as shown in Fig.3a and inset (higher magnification SEM image). Increases in pyrolysis temperature to 1400°C caused the formation of a large amount of nanowires on the cell wall surface, as reported in Fig 3b and the inset (showing the details of nanowires). Instead, pyrolysis at 1250°C under Ar atmosphere resulted in a porous monolith with large cells (>500 µm), having small size pores within the cell wall ant the foam struts, see embedded image in Fig. 3c for strut detail. With increasing the pyrolysis temperature to 1400°C, several cavities (around 20 µm) containing micro-pores appeared on the walls and struts of the ceramic foam. No formation of 1D nanostructures was observed in samples pyrolyzed in Ar (see Fig 3c,d).

It was shown that SiO and CO are the main gaseous species that form during the pyrolysis of a similar polysiloxane precursor at temperatures higher than 1000°C, and that the partial pressure of both of the gases increases with increasing pyrolysis temperature up to 1400°C.[37] In the present study, while the reaction of SiO and $N_2$ governed the formation of $Si_2N_2O$ under $N_2$ pyrolysis, no 1D nanostructure formed during heating under Ar atmosphere. This could be attributable to the differences in nucleation and growth mechanisms of these nanowires; in fact, whereas $Si_2N_2O$ does not necessitate carbon for its formation,[38] a specific level of carbon saturation is needed to precipitate SiC crystals from metal silicides, in another words carbon is crucial for SiC nucleation and growth [39] (note that the precursor, PMS, used in the present study contains a very limited amount of carbon compared to that of other types of polysiloxane precursors commonly used to produce SiOC materials.)[28]. Similarly the explanation for the absence of $SiO_2$ 1D nanostructures could be a higher demand of oxygen saturation compared to that of silicon oxynitride. Consequently, we believe that the lack of SiC or $SiO_2$ nanowires formation under Ar pyrolysis is simply due to the low degree of carbon or oxygen saturation which did not lead a supersaturated Si-C-(O)-Co liquid droplet. It should be noted that when a poly(methyl-phenyl)siloxane was used as a precursor, SiC nanowires formed under the same experimental conditions.[40]

The released SiO gas from PMS precursor reacted with the atmosphere and yielded $Si_2N_2O$ nanowires with tips upon heating in $N_2$ (see detailed image in Fig. 3b). EDS analysis taken from the nanowire caps showed that these tips are comprised of cobalt silicide (data not shown here). A metal-containing cap in the tip of the $Si_2N_2O$ nanowires implies that the nanowires formed by the well-established vapor–liquid–solid (VLS) mechanism. Crystalline silicon oxynitride has been shown to exhibit enhanced excellent refractory properties, compared to silicon nitride.[41] Moreover, it possesses high strength, low thermal expansion, high thermal shock and abrasion resistance, and excellent chemical stability in acid, molten non-ferrous metals and air at high temperature.[42; 43] It was demonstrated that the amorphous $Si_2N_2O$ residue (produced by the pyrolysis of copolymer precursor at 1000°C in $N_2$) can be crystallized at different temperatures to produce $Si_2N_2O$ or $\alpha$-$Si_3N_4$-$Si_2N_2O$ mixtures.[44] A high C:O ratio in the residue resulted in the consumption of the oxygen during crystallization, and $Si_3N_4$ was produced. A low C:O ratio (as in the case of the present study) led to the formation of $Si_2N_2O$. Indeed, similar experiments, performed using a polysiloxane with higher C:O, yielded $Si_3N_4$ without any silicon oxynitride[40] phase (for pyrolysis at 1400°C in $N_2$). The monolith pyrolyzed at 1400°C under $N_2$ had a total porosity of 59.4 vol% (34.0 vol% open porosity). Specific surface area values of the samples investigated in the present study and some other samples made by using different types of precursors are under progress and detailed data will be reported elsewhere[40].

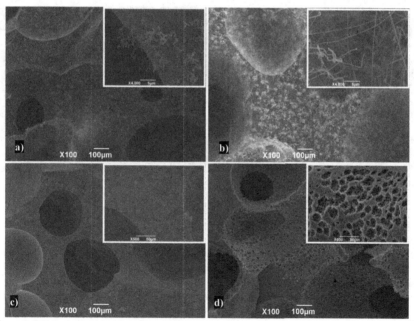

Figure 3. SEM micrographs taken from the fracture surfaces of pyrolyzed PMS-CoCl$_2$-ADA samples treated at: a) 1250°C, b) 1400°C (under N$_2$), and c) 1250°C d) 1400°C (under Ar).

Similar experiments were performed using a polysilazane precursor. SEM images of the fracture surfaces taken from a sample heat treated at 1400°C in Ar are shown in Fig. 4a&b. The micrographs reveal the presence of high degree of interconnected porosity and nanowires on the cell walls and struts, (see Fig. 4a and inset). The specimen possessed an inhomogeneous morphology at the macroscopic scale, as well as some cracks most probably due to the stresses originating from several factors, including the large volume of gas released during pyrolysis. A higher magnification SEM image (Fig. 4b) of the fracture surface reveals that nanowire entanglements were obtained and EDS analysis taken from the tips of these nanowires showed the presence of cobalt silicide phase, implying that the same mechanism for the nucleation and growth of these nanowires, namely, VLS, was active (Au peak comes from the gold sputtering during sample preparation for SEM analysis). Detailed studies are ongoing to extend the research to pyrolysis under $N_2$ and to further characterize these 1D nanostructures protruding from the cell walls. Preliminary XRD investigations, coupled with EDS analysis, indicated that these nanowires are mainly constituted of SiC, although the presence of nitrogen-containing species has not been excluded yet. Comparing the XRD results of precursor treated with and without cobalt (data not shown), it can be said that incorporation of Co catalyst affected the stability of the resulting ceramics at high temperatures by enhancing the crystallization, similarly to what reported for a Ni-containing polysiloxane.[33]

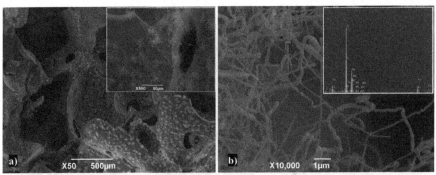

Figure 4. SEM micrographs taken from the fracture surfaces of pyrolyzed VL20-CoCl$_2$-ADA samples treated at 1400°C under Ar. a) low magnification; b) higher magnification (inset: EDS analysis of the tip of a nanowire).

## 4.    CONCLUSIONS

By means of two different preceramic polymers, a blowing agent and a catalyst, macro-porous ceramics with the cell wall surface decorated with 1D nanostructures (nanowires) were produced by catalyst-assisted-pyrolysis. Differences were found depending on the pyrolysis temperature and atmosphere. In Ar, no nanowires formed when using the polysiloxane precursor, presumably because of its low carbon content, while they developed on the surface of a SiCN porous ceramic produced from silazane. Heating in $N_2$ led to the growth of a significant amount of $Si_2N_2O$ nanowires on the surface of the SiOC ceramic, and their amount increased with increasing temperature.

ACKNOWLEDGMENTS

P.C. and C.V. gratefully acknowledge the support of the European Community's Sixth Framework Programme through a Marie-Curie Research Training Network ("PolyCerNet" MRTN-CT-019601).

REFERENCES

1.  F. Schüth, K. S. W. Sing, and J. Weitkamp, "Handbook of porous solids," Wiley-VCH Verlag GmbH Weinheim, Germany. (2002).
2.  M. Twigg and J. T. Richardson, "Fundamentals and Applications of Structured Ceramic Foam Catalysts," *Industrial & Engineering Chemistry Research,* **46** 4166-4177.2007.
3.  Y. Xia, P. Yang, Y. Sun, Y. Wu, B. Mayers, B. Gates, Y. Yin, F. Kim, and H. Yan, "One-Dimensional Nanostructures: Synthesis, Characterization, and Applications," *Advanced Materials,* **15[5]** 353-389.2003.
4.  J. Zheng, M. J. Kramer, and M. Akinc, "In Situ Growth of SiC Whisker in Pyrolyzed Monolithic Mixture of AHPCS and SiC," *Journal of the American Ceramic Society,* **83[12]** 2961-2966.2000.
5.  H. Wang, X.-D. Li, T.-S. Kim, and D.-P. Kim, "Inorganic polymer-derived tubular SiC arrays from sacrificial alumina templates," *Applied Physics Letters,* **86[17]** 173104-173103.2005.
6.  C. Wan, G. Guo, and Q. Zhang, "SiOC ceramic nanotubes of ultrahigh surface area," *Materials Letters,* **62[17-18]** 2776-2778.2008.
7.  H.-M. Yen, S. Jou, and C.-J. Chu, "Si-O-C nanotubes from pyrolyzing polycarbosilane in a mesoporous template," *Materials Science and Engineering: B,* **122[3]** 240-245.2005.
8.  K. F. Cai, Q. Lei, and L. C. Zhang, "Ultra long SiC/SiO2 core-shell nanocables from organic precursor," *Journal of Nanoscience and Nanotechnology,* **5[11]** 1925-1928.2005.
9.  F. Li, G. Wen, and L. Song, "Growth of nanowires from annealing SiBONC nanopowders," *Journal of Crystal Growth,* **290[2]** 466-472.2006.
10. Y. Xu, A. Zangvil, J. Lipowitz, J. A. Rabe, and G. A. Zank, "Microstructure and Microchemistry of Polymer-Derived Crystalline SiC Fibers," *Journal of the American Ceramic Society,* **76[12]** 3034-3040.1993.
11. M. Scheffler, E. Pippel, J. Woltersdorf, and P. Greil, "In situ formation of SiC-Si₂ON₂ micro-composite materials from preceramic polymers," *Materials Chemistry and Physics,* **80[2]** 565-572.2003.
12. D. D. Jayaseelan, W. E. Lee, D. Amutharani, S. Zhang, K. Yoshida, and H. Kita, "In Situ Formation of Silicon Carbide Nanofibers on Cordierite Substrates," *Journal of the American Ceramic Society,* **90[5]** 1603-1606.2007.
13. M. Scheffler, P. Greil, A. Berger, E. Pippel, and J. Woltersdorf, "Nickel-catalyzed in situ formation of carbon nanotubes and turbostratic carbon in polymer-derived ceramics," *Materials Chemistry and Physics,* **84[1]** 131-139.2004.
14. S. Otoishi and Y. Tange, "Effect of a Catalyst on the Formation of SiC Whiskers from Polycarbosilane. Nickel Ferrite as a Catalyst," *Bulletin of the Chemical Society of Japan,* **72[7]** 1607-1613.1999.
15. W. Yang, H. Miao, Z. Xie, L. Zhang, and L. An, "Synthesis of silicon carbide nanorods by catalyst-assisted pyrolysis of polymeric precursor," *Chemical Physics Letters,* **383[5-6]** 441-444.2004.
16. A. Berger, E. Pippel, J. Woltersdorf, M. Scheffler, P. Cromme, and P. Greil, "Nanoprocesses in polymer-derived Si-O-C ceramics: Electronmicroscopic observations and reaction kinetics," *physica status solidi (a),* **202[12]** 2277-2286.2005.
17. J. Haberecht, F. Krumeich, M. Stalder, and R. Nesper, "Carbon nanostructures on high-temperature ceramics - a novel composite material and its functionalization," *Catalysis Today,* **102-103** 40-44.2005.
18. W. Yang, Z. Xie, J. Li, H. Miao, L. Zhang, and L. An, "Ultra-Long Single-Crystalline α-Si3N4 Nanowires: Derived from a Polymeric Precursor," *Journal of the American Ceramic Society,* **88[6]** 1647-1650.2005.

19. K. F. Cai, L. Y. Huang, A. X. Zhang, J. L. Yin, and H. Liu, "Ultra Long SiCN Nanowires and SiCN/SiO2 Nanocables: Synthesis, Characterization, and Electrical Property," *Journal of Nanoscience and Nanotechnology,* **8** 6338-6343.2008.
20. W. Yang, H. Wang, S. Liu, Z. Xie, and L. An, "Controlled Al-Doped Single-Crystalline Silicon Nitride Nanowires Synthesized via Pyrolysis of Polymer Precursors," *The Journal of Physical Chemistry B,* **111**[16] 4156-4160.2007.
21. W. Yang, F. Gao, H. Wang, X. Zheng, Z. Xie, and L. An, "Synthesis of Ceramic Nanocomposite Powders with in situ Formation of Nanowires/Nanobelts," *Journal of the American Ceramic Society,* **91**[4] 1312-1315.2008.
22. B.-H. Yoon, C.-S. Park, H.-E. Kim, and Y.-H. Koh, "In Situ Synthesis of Porous Silicon Carbide (SiC) Ceramics Decorated with SiC Nanowires," *Journal of the American Ceramic Society,* **90**[12] 3759-3766.2007.
23. C. Vakifahmetoglu, E. Pippel, J. Woltersdorf, and P. Colombo, "Growth of 1D-Nanostructures in Porous Polymer Derived Ceramics by Catalyst-Assisted-Pyrolysis. Part I: Iron Catalyst," *submitted to Journal of the American Ceramic Society.*2009.
24. C. Vakifahmetoglu, I. Menapace, A. Hirsch, L. Biasetto, R. Hauser, R. Riedel, and P. Colombo, "Highly Porous Macro- and Micro-Cellular Ceramics from a Polysilazane Precursor," *Accepted for publication in Ceramics International.*2009.
25. ASTM D 3576, "Standard test method for cell size of rigid cellular plastics." In *Annual Book of ASTM Standards, Vol. 08.02.* West Conshohocken, PA., 1997.
26. T. Takahashi and P. Colombo, "SiOC Ceramic Foams through Melt Foaming of a Methylsilicone Preceramic Polymer," *Journal of Porous Materials,* **10** 113-121.2003.
27. C. Vakifahmetoglu and P. Colombo, "A Direct Method for the Fabrication of Macro-Porous SiOC Ceramics from Preceramic Polymers," *Advanced Engineering Materials,* **10**[3] 256-259.2008.
28. M. Scheffler, T. Gambaryan-Roisman, T. Takahashi, J. Kaschta, H. Muenstedt, P. Buhler, and P. Greil, "Pyrolytic decomposition of preceramic organo polysiloxanes," *Innovative processing and synthesis of ceramics, Glasses Compos IV: Ceramic Transactions,* **115** 239–250.2000.
29. C. G. Pantano, A. K. Singh, and H. Zhang, "Silicon Oxycarbide Glasses," *Journal of Sol-Gel Science and Technology,* **14**[1] 7-25.1999.
30. G. D. Soraru, S. Modena, E. Guadagnino, P. Colombo, J. Egan, and C. Pantano, "Chemical Durability of Silicon Oxycarbide Glasses," *Journal of the American Ceramic Society,* **85**[6] 1529-1536.2002.
31. D. Walter and I. W. Karyasa, "Synthesis and Characterization of Cobalt Monosilicide (CoSi) with CsCl Structure Stabilized by a beta-SiC Matrix," *Zeitschrift für anorganische und allgemeine Chemie,* **631**[6-7] 1285-1288.2005.
32. H. J. Whitlow, Y. Zhang, C. M. Wang, D. E. McCready, T. Zhang, and Y. Wu, "Formation of cobalt silicide from filter metal vacuum arc deposited films," *Nuclear Instruments and Methods in Physics Research Section B: Beam Interactions with Materials and Atoms,* **247**[2] 271-278.2006.
33. M. G. Segatelli, A. T. N. Pires, and I. V. P. Yoshida, "Synthesis and structural characterization of carbon-rich SiCxOy derived from a Ni-containing hybrid polymer," *Journal of the European Ceramic Society,* **28**[11] 2247-2257.2008.
34. P. Colombo, M. O. Abdirashid, M. Guglielmi, L. Mancinelli Degli Esposti, and A. Luca, "Preparation of Ceramic composites by active-filler-controlled-polymer-pyrolysis," Materials Research Society Symposia Proceedings, **346** 403-408.1994.
35. J. Zeschky, T. Höfner, C. Arnold, R. Weißmann, D. Bahloul-Hourlier, M. Scheffler, and P. Greil, "Polysilsesquioxane derived ceramic foams with gradient porosity," *Acta Materialia,* **53**[4] 927-937.2005.
36. S. Quinn, "Chemical blowing agents: providing production, economic and physical improvements to a wide range of polymers," *Plastics, Additives and Compounding,* **3** 16-21.2001.
37. Q. Wei, E. Pippel, J. Woltersdorf, M. Scheffler, and P. Greil, "Interfacial SiC formation in polysiloxane-derived Si-O-C ceramics," *Materials Chemistry and Physics,* **73**[2-3] 281-289.2002.

38.  J. Zheng, X. Song, X. Li, and Y. Pu, "Large-Scale Production of Amorphous Silicon Oxynitride Nanowires by Nickel-Catalyzed Transformation of Silicon Wafers in NH3 Plasma," *The Journal of Physical Chemistry C,* **112**[1] 27-34.2008.
39.  G. Yang, R. Wu, M. Gao, J. Chen, and Y. Pan, "SiC crystal growth from transition metal silicide fluxes," *Crystal Research and Technology,* **42**[5] 445-450.2007.
40.  C. Vakifahmetoglu, S. Carturan, E. Pippel, J. Woltersdorf, and P. Colombo, "Growth of 1D-Nanostructures in Porous Polymer Derived Ceramics by Catalyst-Assisted-Pyrolysis. Part II: Cobalt Catalyst," *manuscript in preparation.*2009.
41.  V. Weeren, E. A. Leone, S. Curran, L. C. Klein, and S. C. Danforth, "Synthesis and Characterization of Amorphous $Si_2N_2O$," *Journal of the American Ceramic Society,* **77**[10] 2699-2702.1994.
42.  G. Chollon, U. Vogt, and K. Berroth, "Processing and characterization of an amorphous Si–N–(O) fibre," *Journal of Materials Science,* **33**[6] 1529-1540.1998.
43.  M. E. Washburn, "Silicon oxynitride refractories," *Amer. Ceram. Soc. Bull.,* **46** 667-671.1967.
44.  G.-E. Yu, J. Parrick, M. Edirisinghe, D. Finch, and B. Ralph, "Synthesis of silicon oxynitride from a polymeric precursor," *Journal of Materials Science,* **28**[15] 4250-4254.1993.

SYNTHESIS OF CERAMIC NANO FIBER FROM PRECURSOR POLYMER BY SINGLE
PARTICLE NANO-FABRICATION TECHNIQUE

Masaki Sugimoto, Akira Idesaki, Masahito Yoshikawa
Quantum Beam Science Directorate, Japan Atomic Energy Agency,
Takasaki, Gunma, Japan

Shogo Watanabe, Shu Seki
Department of Applied Chemistry, Osaka University,
Suita, Osaka, Japan

ABSTRACT
    We have succeeded in synthesizing palladium containing ceramic nano fibers from precursor polymers using Single Particle Nanofabrication Technique (SPNT). A thin film of the precursor polymer was cross-linked by MeV-order heavy-ion beam irradiation along ion tracks of nano-size in radius through the whole thickness of the thin film. Nano fibers were developed on the surface of the substrate by dissolution and washing away of un-crosslinked polymer. Subsequent pyrolysis converted polymeric into ceramic nano fibers. The radius, length and number density of nano fibers are controlled by liner energy transfer (LET) of ion beam, thickness of polymeric target and number of projected ions, respectively.

INTRODUCTION
    Synthesis process of silicon carbide (SiC) ceramics from Polycarbosilane (PCS) was invented by Yajima et al., [1] and continuous SiC ceramic fiber was commercialized as Nicalon supplied from Nippon Carbon Co., Ltd. In the synthesis process, PCS fiber was crosslinked to hold the fiber shape and increase the ceramic conversion yield, and then the cured fiber is able to convert in SiC ceramics. We had succeeded in improving the heat resistance of SiC fiber reducing the oxygen content in the fiber using electron beam crosslinking technique [2, 3]. On the other hand, SiC nanostructures have been shown to exhibit more superior properties than bulk SiC [4]. In addition, the electron field-emission properties of SiC nano fiber show a threshold electric field comparable to that of a carbon nano tube based material. Several techniques have already been developed for synthesizing SiC in the form of nano spheres and nano fibers / rods [4, 5].
    Recently we reported the radiation effects of ion beams on polysilanes and the dependence of reaction processes on linear energy transfer (LET: energy deposition rate of incident particles per unit length) of the characteristic radiation [6], subsequently we succeeded in synthesizing nano fiber by the crosslinking reactions of polymer molecules within an ion track along a particle track [7,8,9]. The present nano-scaled negative tone imaging technique (single particle nano-fabrication technique: SPNT) shows a striking contract to the conventional nuclear track technique, providing the direct formation of nano fibers based on a variety of polymeric materials with fairly controlled sizes (length, radius, number density, etc.). Thus nano fibers reflecting the properties of the target polymer materials have been successfully realized by the simple procedure. Therefore we had also successfully produced SiC ceramic nano fiber by applying SPNT to the PCS [10, 11]. This process suggests that a catalyst-loaded SiC nano fiber as well as SiC material with various shapes can be fabricated if a precursor composed of PCS and transition metal compound is synthesized. In this research, we applied SPNT for ceramic precursor polymer containing metals to synthesis of the SiC ceramic nano fiber, and discussed synthesis mechanism and control method of the length, radius and number density of the fiber.

EXPERIMENTAL PROCEDURE

Synthesis process of ceramic nano fiber by ion beam irradiation is shown in Figure 1. Polycarbosilane (PCS) was used for a precursor polymer. PCS is a solid polymer at ambient temperature and has a number average molecular weight of $2.0 \times 10^3$. Palladium(II) acetate (Pd(OAc)$_2$) was mixed with PCS in order to fold catalyst metal into precursor polymer. Pd(OAc)$_2$ is orange-brown powder with molar mass of 224.5g/mol. Each polymers were dissolved in tetrahydrofuran (THF) separately, then the Pd(OAc)$_2$ solution was added by dropping to the PCS solution with stirring at ambient temperature. After the mixing, the THF was evaporated, and then a polymer blend of PCS and Pd(OAc)$_2$ (PCS-Pd(OAc)$_2$) was obtained. The mass ratio of PCS / Pd(OAc)$_2$ was 15 / 1.

The each polymers solved into toluene, and the solution at 5 mass% was spin-coated on polished Si substrate to make the polymer thin films. The polymer thin films were irradiated in vacuum at ambient temperature using 388 MeV $^{58}$Ni$^{15+}$, 450 MeV $^{129}$Xe$^{23+}$ and 500 MeV $^{197}$Au$^{31+}$ ion beams from cyclotron accelerator at Takasaki Advanced Radiation Research Institute (TIARA), Japan Atomic Energy Agency (JAEA). The loss of kinetic energy of ions due to penetration through the polymer films was estimated using the SRIM 2008 calculation code. In this process, ion irradiation can release densely active intermediates within a cylindrical area along the passage of a single ion. The cylindrical area is sometimes called an "ion track". These intermediates with an inhomogeneous spatial distribution in the ion track promote the various chemical reactions within the area. Ion irradiation at low fluence without overlapping between ion tracks can be recognized each ion track as single ion event in the target materials. The cross-linking reactions along the ion track result in the formation of a nano fiber in thin films. A uncross linked area can be removed by development with organic solvents, utilizing the change in solubility due to the crosslinking of PCS. After irradiation, the samples were treated by toluene for 120 sec. The irradiated part of the film, insoluble in the solvent was developed as polymer nano fibers. This was fired in argon atmosphere at 1273 K for 1800 sec using an electric furnace with a heating ramp rate of 250 K/h and then cooled down to room temperature, giving the final ceramic nano fibers on the substrate. Direct observation of the nano fibers was conducted using Scanning Probe Microscopy (SPM) Seiko Instruments Inc. SPA-400 using dynamic force microscope method. The size of cross-section of nano fiber is defined as an average radius of cross-sectional measurements of a nano fiber.

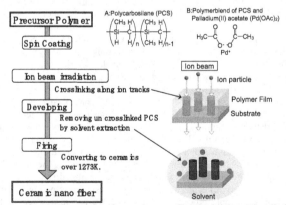

Figure 1. Synthesis process of ceramic nano fiber by ion beam irradiation.

RESULTS AND DISCUSSION

The isolated nano fibers of each precursor polymers which formed by ion beam irradiation on the Si substrate are observed directly by SPM micrographs. These polymer nano fibers were fired at 1273 K in argon. The surface of the substrate was observed using the SPM again, as shown in Figure 2. The nano fibers remained on the substrate after firing, and the shape did not change by heating again over 1573 K in argon and 1273 K in air. This result indicates that the fired nano fiber has good thermal and oxidation resistance and it was converted into ceramics via an organic to inorganic conversion reaction similar to the Nicalon fiber.

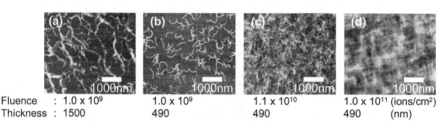

| Fluence | : 1.0 x 10$^9$ | 1.0 x 10$^9$ | 1.1 x 10$^{10}$ | 1.0 x 10$^{11}$ (ions/cm²) |
| Thickness | : 1500 | 490 | 490 | 490 (nm) |

Figure 2. SPM micrographs of ceramic nano fibers formed from the thin film of PCS-Pd(OAc)$_2$ by 450 MeV Xe ion beam irradiation.

The measured average length of the fired nano fibers in Figure 2 (a) and (b) were 1470 and 480 nm respectively, the lengths were almost the same as thickness of each thin films. The nano fibers fell down on the substrate and combined with the surface of substrate partially, so the fiber shrank with precedence in radius direction and did not shrink in length direction during firing. The length of the polymer nano fiber can be controlled by the thickness of coated polymer [7], therefore the length of ceramic nano fiber also controlled by changing the film thickness within the range of penetration length of the projected ions. Furthermore, Figure 2 (b), (c) and (d) shows the number densities of ceramic nano fiber were increasing with fluence of ion beam irradiation. Figure 3 demonstrates the changes in the number density of isolated polymer nano fiber formed from PCS and PCS-Pd(OAc)$_2$ thin films with 490 nm thick on the Si substrate by 450 MeV Xe ion irradiation with various fluence. In addition, the number density of ceramic nano fiber after firing at 1273 K are plotted versus fluence in Figure 3, showing, striking consistency with the fluence of irradiated ions. This also suggests that each nano fiber is produced within a single ion track. In the range of fluence over $1 \times 10^{10}$ (ions/cm₂), the measured number of density has decreased compared with the theoretical values. This reason seems to be same as a past report that overlapping of the tracks becomes significant in the fluence range, leading to underestimate for the number of nano fibers on 2D substrate [6]. The number density of fired nano fiber is quite the same as that of polymer nano fiber, suggesting corresponding conversion of polymer nano fiber into ceramics without detachment from substrate during firing.

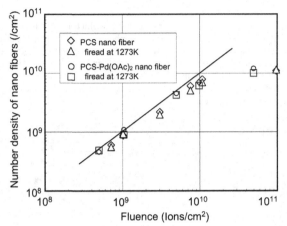

Figure 3. The number density of nano fibers formed from PCS and PCS-Pd(OAc)₂ thin films on the Si substrate by 450 MeV Xe ion irradiation with various fluence.

Radii of polymeric and fired nano fibers formed by ion beam irradiation with various LET are presented in Figure 4.   The radius depended on the radial distribution of energy in a particle track, and the cylindrical area of energy deposition can be divided into two parts: core and penumbra, and the following formulae already exist for the values of the coaxial energy in an ion track: [12]

$$\rho_c(r) = \frac{LET}{2}[\pi r^2]^{-1} + \frac{LET}{2}\left[2\pi r_c^2 \ln\left(\frac{e^{1/2}r_p}{r_c}\right)\right]^{-1} \quad r \leq r_c \tag{1}$$

$$\rho_p(r) = \frac{LET}{2}\left[2\pi r^2 \ln\left(\frac{e^{1/2}r_p}{r_c}\right)\right]^{-1} \quad r_c < r \leq r_p \tag{2}$$

where $\rho_c(r)$ and $\rho_p(r)$ are the deposited energy density at core and penumbra area, respectively; e is an exponential factor; $r_c$ and $r_p$ are the radii of core and penumbra area.   In the case of PCS, $r_c$ of 500 MeV Au irradiation estimated by SRIM code are 0.74 nm, and this values are enough smaller than radius of each PCS nano fibers.   In this case, the radius of the nano fibers are depend on only the energy distribution in the penumbra and can be described by the equation (2) [13, 14].   For gel formation in a polymer system, it is necessary to introduce one crosslink per polymer molecule. Assuming a sole contribution from the crosslinking reactions in the chemical core, $\rho_{cr}$ is given by

$$\rho_{cr} = \frac{100\rho A}{G(x)mN} \tag{3}$$

where $A$ is Avogadro's number, $m$ is the mass of the monomer unit, and $N$ is the degree of polymerization [8]. $G(x)$ denotes the efficiency of cross-linking reaction: number of cross-links produced by 100 eV absorbed by the polymer materials. Substitution of $\rho_p$ in Eq. (2) with $\rho_{cr}$ gives the following requirement for $r_{cr}$ :

$$r_{cr}^2 = \frac{LET \cdot G(x)mN}{400\pi\rho A}\left[\ln\left(\frac{e^{1/2}r_p}{r_c}\right)\right]^{-1} \tag{4}$$

we had reported that Eq.(4) is corresponding to the experimental result of the PCS nano fiber [10, 11] and several kinds of polymers well [6, 9], and Eq.(4) also gives good interpretation to the PCS-Pd(OAc)$_2$ in Figure 4. However, the radii of the PCS-Pd(OAc)$_2$ nano fiber are a little larger than that of the PCS nano fiber. This reason is thought to be related to the difference between G(x), because the catalytic Pd in the PCS enhances the crosslinking reaction. By SEM and EDS observation of fired samples, there is no Pd crystal on the substrate surface, however the Pd element was observed. This is indicate that PCS-Pd(OAc)2 nano fiber could fired without elution and aggregation, then Pd is remain in the nano fiber after firing.

The polymer nano fibers seems to shrink by firing in a constant ratio as shown in Figure 4, accordingly the shrinkage ratio were calculated by equation (5);

$$\text{Shrinkage ratio} = (r_{SiC})^2 / (r_{PCS})^2 \qquad (5)$$

where $r_{SiC}$ and $r_{PCS}$ is radius of the fired SiC nano fiber and PCS nano fiber respectively shown in Fig. 4. The PCS nano fibers mainly shrank in the radius direction and the length of the fiber almost does not change during firing as shown in Fig. 2, therefore the shrinkage ratio is derived from change of the radius of the fibers. The shrinkage ratio of the nano fibers was distributed within the range from 0.38 to 0.42; these values are a little large compared with that of SiC fiber such as Nicalon® fiber with micron order diameter. It is reported that the ion beam irradiation at doses ranging over $1\times10^{14}$ ions/cm$^2$ induce ceramization with a novel mechanism different from thermal pyrolysis, then the precursor film convert into the inorganic ceramic film. [15, 16] Similarly for the SPNT, the ion beam gives the very high energy to the core part of the polymer nano fiber, and the core part may change into ceramics during irradiation, therefore it is thought that the shrinkage ratio became small compared with that of Nicalon® fiber.

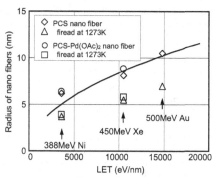

Figure 4. Radii of polymer nano fibers formed by ion beam irradiation with various LET determined by the cross section profile of SPM micrograph.

CONCLUSION

We have succeeded in synthesizing palladium containing ceramic nano fibers from Si based precursor polymers using Single Particle Nanofabrication Technique (SPNT) by MeV-order heavy-ion beam irradiation. Non-homogeneous field of chemical reactions in an ion track leads cylindrical nano structures based on crosslinking reactions of polymeric materials. The nano fibers of polymer blend with PCS and Pd(OAc)$_2$ were developed on the surface of the Si substrate by dissolution and can convert into ceramic nano fiber by firing. The sizes of the fiber are completely under control radius, length and number density. SPNT will enables to fabricate nano fibers with supreme large surface area from many kind of polymers containing catalyst elements, therefore it has been expected as high efficiency catalyst and adsorption materials.

ACKNOWLEDGMENT
This work was supported by a Grant-in-Aid for Scientific Research from the Japan Society for the Promotion of Science.

REFERENCES
[1] S. Yajima, J. Hayashi, M. Omori, Continuous silicon carbide fiber of high tensile strength, *Chem. Lett.*, **9**, 931-934 (1975).
[2] K. Okamura, T. Seguchi, Application of radiation curing in the preparation of polycarbosilane-derived SiC fibers, *J. Inorg. Organometal. Polym.*, **1**, 171-179 (1992).
[3] M. Sugimoto, T. Shimoo, K. Okamura, and T. Seguchi, Reaction Mechanisms of Silicon Carbide Fiber Synthesis by Heat Treatment of Polycarbosilane Fibers Cured by Radiation, *J. Am. Ceram. Soc.*, **78**, 1013-17 (1995)
[4] E.W. Wong, P.E. Sheehan, C.M. Lieber, Nanobeam Mechanics: Elasticity, Strength, and Toughness of Nanorods and Nanotubes, *Science,* **277**, 1971-1975 (1997).
[5] G. Shen, D. Chen, K. Tang, Y. Qian and S. Zhang, Silicon carbide hollow nanospheres, nanowires and coaxial nanowires, *Chem. Phys. Lett.*, **375**, 177-184 (2003).
[6] S. Seki, K. Maeda, Y. Kunimi, S. Tagawa, Y. Yoshida, H. Kudoh, M. Sugimoto, Y. Morita, T. Seguchi, T. Iwai, H. Shibata, K. Asai and K. Ishigure, Ion Beam Induced Crosslinking Reactions in Poly(di-n-hexylsilane), *J. Phys. Chem. B*, **103**, 3043-3048 (1999).
[7] S. Seki, K. Maeda, S. Tagawa, H. Kudoh, M. Sugimoto, Y. Morita and H. Shibata, Formation of Nanowires along Ion Trajectories in Si Backbone Polymers, *Adv. Mater.*, **13**, 1663-1665 (2001)
[8] S. Seki, S. Tsukuda, K. Maeda, Y. Matsui, A. Saeki and S. Tagawa, Inhomogeneous distribution of crosslinks in ion tracks in polystyrene and polysilanes, *Phys. Rev. B*, **70**, 144203 (2004).
[9] S. Seki, S. Tsukuda, K. Maeda, S. Tagawa, H. Shibata, M. Sugimoto, K. Jimbo, I. Hashitomi and A. Koyama, *Macromolecules*, **38**, 10164 (2005).
[10] S. Tsukuda, S. Seki, S. Tagawa, M. Sugimoto, A. Idesaki, S. Tanaka and A. Ohshima, Fabrication of Nanowires Using High-Energy Ion Beams, *J. Phys. Chem.*, **108**, 3407-3409 (2004).
[11] M. Sugimoto, M. Yoshikawa, S. Tsukuda and S. Seki, Synthesis of Ceramic Nano fiber from Precursor Polymers by Ion Beam Irradiation, *Transac. Mater. Res. Soc. Jpn.*, **33** 1027-1030 (2008).
[12] J. L. Magee, A. Chatterjee, Kinetics of Nonhomogenous Processes, Ed. by G.R. Freeman, Wiley, New York, 171 (1987).
[13] J. L. Magee and A. Chatterjee, Radiation chemistry of heavy-particle tracks. 1. General considerations, *J. Phys. Chem.* **84**, 3529-3536 (1980).
[14] A. Chatterjee and J.L. Magee, Radiation chemistry of heavy-particle tracks. 2. Fricke dosimeter system, *J. Phys. Chem.* **84**, 3537-3543 (1980).
[15] J. C. Pivin, P. Colombo and M. Tonidandel, Ion Irradiation of Preceramic Polymer Thin Films, *J. Am. Ceram. Soc.*, **79**, 1767-70 (1996).
[16] J. C. Pivin and P. Colombo, Ceramic coatings by ion irradiation of polycarbosilanes and polysiloxanes - Part I Conversion mechanism, *J. mater. Sci.*, **32**, 6163-6173 (1997).

# SYNTHESIS OF NOVEL SiBNC FIBER PRECURSOR BY A ONE-POT ROUTE

Yun Tang, Jun Wang, Xiao-dong Li, Yi Wang

*State Key Laboratory of Advanced Ceramic Fibers & Composites, National University of Defense Technology, Changsha, P. R. China.*

ABSTRACT

A novel processable N-methylpolyborosilazane was synthesised by a one-pot method by using boron trichloride (BTC), trichlorosilane (TCS) and heptamethyldisilazane ($H_p$MDZ) as the starting materials. The reaction mainly involves the condensation between B-Cl, Si-Cl and $-SiMe_3$ with $ClSiMe_3$ evaporation. The steric $N-CH_3$ in the $H_p$MDZ plays an important role in improving the processing properties of the polymer. The obtained polymer has a relatively high softening point and can easily be melt-spun into polymer fibers and then be converted into high performances SiBNC fibers. This route is suitable for the continuous preparation of the preceramic polymer for SiBNC fibers under mild conditions.

INTRODUCTION

Because of their lightweight and excellent thermal and oxidative stabilities, ceramic fibers composed of Si, B, N and C are suitable candidates as reinforcement in ceramic matrix composites for high temperature applications.[1-4] However, the lack of general synthesis routes which allow for the generation of these materials with controlled composition and under moderate conditions, has hampered both scientific studies and their practical applications. Polymer pyrolysis is the only effective approach known for the preparation of SiBNC fibers.

Initial work in the preparation of preceramic polymers for SiBNC fibers was done by Takamizawa et al[5] by heating a mixture of organopolysilane and organoborazine in an inert atmosphere. Sneddon et al[6-8] synthesised polyborosilazanes with appropriate rheological properties through chemical modification of hydridopolysilazanes with various borane or borazine derivatives. A very nice job for the fabrication of SiBNC fibers was done by Jansen et al and a pilot plant has been successfully built for the production of such continuous fibers with high performances[3, 9-11]. They synthesised the polymer N-methylpolyborosilazanes by aminolysis of single source precursor $Cl_3Si-NH-BCl_2$ or $MeCl_2Si-NH-BCl_2$. Starting from another single source precursor $B(C_2H_4SiCH_3Cl_2)_3$, Bernard et al[2, 12] obtained the $[B(C_2H_4SiCH_3NCH_3)_3]_n$ precursor which can also be used as the preceramic polymer to SiBNC fiber.

Obviously, these synthesis routes for the polymers are multiple-step processes, that is, before the polymerization of the target polymeric precursor, additional steps are required to prepare the reactant oligomers or the single source precursors. Such molecular or polymeric intermediates are usually extremely sensitive to moisture and the generation of sufficiently inert and dry reaction environment can be very difficult[13] . Moreover, processes such as the removal of the byproducts are also needed separately.

An alternative access to SiBNC precursors was developed by Lee et al [14]. They prepared a polyborosilazane polymer by reacting boron trichloride (BTC), trichlorosilane (TCS) with hexamethyldisilazane ($H_x$MDZ). The process proves to be simpler and cheaper since no additional cross-linking agents were needed and the byproducts can be removed directly. In this route, the polymerization proceeds quickly and provides highly cross-linked polymers, making them unsuitable for fiber fabrication. One of the important reasons for the quick polymerization and therefore the difficulty to get polymers with appropriate processablities lies in the dehydrogenation of Si-H and N-H which occurs readily under moderate conditions. Practically, in Lee's route, we used dichloromethylsilane (DCMS) instead of TCS as one of the starting monomers, and the resulted polymer always has a low softening point (~70 °C) which is hard to increase due to the readily dehydrogenation reaction of Si-H and N-H.[15] Accordingly, the comparison of different SiBNC fiber precursor points to the fact that the Si-H containing monomer and N-H containing monomer are not utilized at the same time as the starting material in the synthesis of high softening point polymers. As is known, appropriate rheology with suitable melt-processability is one of the important prerequisites for fiber precursors. The key to such a property is to eliminate or inhibit the polymer cross-linking reactions. One way to control the dehydrogenation reactions that result in cross-linking is to remove the Si-H groups or N-H groups in the starting monomers.

This paper reports i) the synthesis by a original one-pot route of a novel processable polyborosilazane using boron trichloride, trichlorosilane and heptamethyldisilazane (H$_p$MDZ) which contains the N-CH$_3$ group as the starting monomers and ii) the first attempts at producing SiBNC fibers from this new polymeric precursor. This mild one-pot chemical process developed here is well suited for the continuous preparation of a low cross-linked polyborosilazane. To the best of our knowledge, it is the first time that a mild one-pot route has been proposed for the synthesis of a spinnable N-methylpolyborosilazane which could also permit the formation of high performances SiBNC fibers.

EXPERIMENTAL PROCEDURE

All reactions were carried out in a purified nitrogen atmosphere using Schlenk-type techniques as described by Shriver[16]. All of the monomers were handled without air contact and stored in a moisture free environment. In a typical reaction, BTC (11.7g, 0.1mol) in n-hexane (100ml) solution and TCS (13.5g, 0.1mol) were introduced and mixed into a pre-cooled reactor. H$_p$MDZ (70.2g, 0.4mol) was introduced into a dropping funnel and added dropwisely to the mixed solution under vigorous stirring. The reactor was kept below 0 °C during the whole addition. After the addition, the reaction mixture was heated slowly to 200~320 °C and held at the final temperature for 6~20 hour. The product (15.3g) was collected as light yellow transparent bulky solid after a vacuum evaporation. The yield of product was 91% of theory, based on the weight of BTC.

Polymer fibers were prepared using a lab-scale melt-spinning system set up inside a nitrogen-filled glove-box. The as-synthesized polymer was fed into an extruder, where it is heated, sheared, and pressured through a filtering system to eliminate any gels or unmelts that may be present in the extrudate. The molten polymer then passed through a spinneret that has a single 0.25 mm capillary. The extrudate flow was then uniaxially drawn in nitrogen atmosphere to filament, which was subsequently stretched and collected on a rotating spool.

Fourier transform infrared (FTIR) spectra were obtained with a Nicolet Avatar 360 in a KBr pellet. $^1$H-, $^{11}$B- and $^{29}$Si-nuclear magnetic resonance (NMR) spectra were performed in CDCl$_3$ with a Bruker Avance 400. Tetramethylsilane (TMS) was used as an internal standard for $^1$H-NMR and $^{29}$Si-NMR. The chemical shift of $^{11}$B-NMR was referenced to BF$_3$·OEt$_2$. Quantitative analyses of nitrogen have been carried out in a Leco TC–436 N/S Determinator, whereas carbon was measured in an Elemntar Vario EL, ELTRA CS–444 C/S analyzer. Silicon and boron were quantified by means of ICP–AES in an Arl 3580B spectrometer after digestion of the ceramic samples using a mixture of bases. O was measured in an IRO–I Oxygen Determinator. Thermal properties (polymer softening and decomposition) were studied by differential scanning calorimetry (DSC CD2–34P) in a nitrogen atmosphere between –50 and 400 °C at a heating rate 10 °C min$^{-1}$ in alumina crucibles. Scanning electron microscopy (SEM) images of green fibers were taken with a JSW5600LV. The obtained pyrolyzed specimens were characterized by powder X-ray diffraction (XRD, D8 ADVANCEX), using Cu–K$_a$ radiation. Tensile strengths of the as-obtained fibers were determined from failure tests performed on 30 filaments with a gauge length of 25mm by using the statistical approach of Weibull. Young's modulus was evaluated from the strength strain curves.

RESULTS AND DISCUSSION

Typically, the polymer was prepared by condensation of BTC, TCS and H$_p$MDZ in a molar ratio of 1:1:4 in n-hexane, in which excess of H$_p$MDZ was used to limit the molecular weight of the polymer. The Me$_3$SiCl and H$_p$MDZ formed can be removed by in situ distillation. The proposed reaction scheme and polymer structure were illustrated in equation 1.

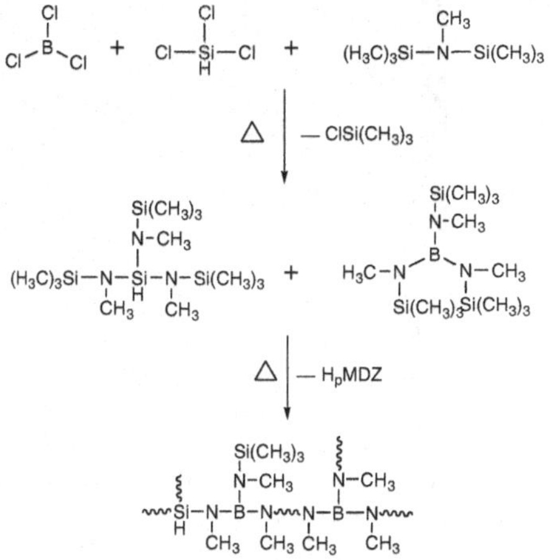

Equation 1

Element analysis* of the polymer is consistent with the formula Si$_{1.1}$BN$_{3.2}$C$_{3.6}$H$_{12}$**, while an N/B ratio of 4 is not expected since the loss of H$_p$MDZ occurs during the formation of polymeric networks. The Si/B ratio is more than 1 since some SiMe$_3$ groups exists as pendant or end groups. Considering the N/C ratio and the steric hindrance of N-CH$_3$ groups, we can logically assume that the reaction of N-CH$_3$ with Si-H is just a minor pathway if can not be excluded. Accordingly, liquid-state $^1$H, $^{11}$B and $^{29}$Si-NMR§ spectra support the proposed polymer structure. We did not find any peak attributed to B-Cl bonds in the IR and $^{11}$B-NMR spectrum. Consistently, no chlorine was detected by elemental analysis. The IR data indicate the formation of B-N bonds since the spectrum shows bands at

1478, 1382 and 1340 cm$^{-1}$ [17-20]. Moreover, the $^{11}$B-NMR spectrum shows singlet resonances at 28.2 ppm, confirming the BN$_3$ units in the polymer structure* [20-22].

This novel polymer has a softening point* of ~ 143 °C and exhibits good rheological properties for melt-spinning. The relative high softening point facilitates the subsequent processes such as the curing of the green fibers and the final ceramic fibers therefrom always show higher mechanical performance[17]. Using a lab scale melt-spinning apparatus which was set up in a nitrogen-filled glove-box, continuous green fibers with > 600 m in length and ~15 µm in diameter can be easily obtained at 172 °C under conditions without optimization. The resulted fibers are smooth, uniform and free of defects (Fig.1 and Fig. 2). The good rheological property mirrors the high chain flexibility resulting from the low cross-linking degree of the polymeric networks. In contrast, in Lee's route in which H$_x$MDZ was used, while using the same monomer ratio and at a lower temperature, the resulted polymer was highly cross-linked. Therefore, we speculate that the steric hindrance of N-CH$_3$ groups in the H$_p$MDZ plays a significant role in reducing the cross-linking degree of the polymer.

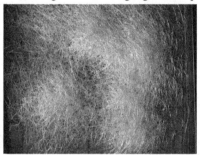

Fig. 1 Optic images of polyborosilazane fibers

Prior to pyrolysis, chemical curing with vapor BCl$_3$ was utilized to make the green fibers infusible at temperature > 60 °C. The -SiMe$_3$ groups in the polymer react readily with B-Cl with the formation of volatile ClSiMe$_3$. Subsequent pyrolysis of the sufficiently cured fibers was conducted in N$_2$ in the same furnace up to 1400 °C yielding fibers with the composition SiBN$_{2.8}$C$_2$ **by element analysis ***. Being amorphous, the SiBNC fibers have smooth surfaces as shown by SEM images (Fig. 2). Although crystalline phases couldn't be identified in the X-ray diffraction patterns, amorphous boron nitride, silicon nitride and silicon carbon phases were clearly identified in the as-obtained SiBNC fibers by the corresponding characteristic absorption bands in the FT-IR spectrum. Typically, the fibers are ~ 10 µm in diameter retaining their circular shape without inter-fusion during the curing and pyrolysis. Moreover, the fibers exhibit good mechanical strength with the tensile strengths ~ 2.2 GPa and elastic modulus ~ 230 GPa. Improvement of these values can be expected after optimization of all the procedures.

CONCLUSION

In summary, this study demonstrates the possibility of fabricating high performances SiBNC fibers starting from an original polyborosilazane obtained by a one-pot method. This route is particularly interesting since the preparation of the target polymer can be achieved in a single reactor. It contains Si-N-B flexible bridges with blocking -N-CH$_3$ groups. The polymer has a relatively high softening point which favours the subsequent processing of the fibers. The as-obtained SiBNC fibers show good mechanical properties. Moreover, considerable increase in the fibers' mechanical properties can be expected by improving each step of this process.

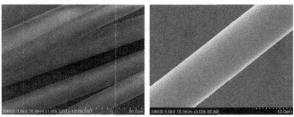

Fig. 2 SEM images of green fiber (a) and as-obtained SiBNC fiber (b)

FOOTNOTES

*Anal. found. (wt %) Si: 22.03; B: 6.44; N: 32.05; C: 30.9; H: 8.58. DSC: $T_m$ ~143 °C. NMR: the polymer was dissolved in $CDCl_3$, $H_3BO_3$ and $Si(CH_3)_4$ was used as external and internal reference, respectively. The [11]B-NMR chemical shift are referenced to $Et_2O \cdot BF_3$ ([11]B, $\delta = 0$) and [1]H, [29]Si-NMR chemical shift are referenced to $Si(CH_3)_4$ ([1]H, [29]Si, $\delta = 0$). [11]B ($\delta$ ppm) ($CDCl_3$): 28.2 (s, B $(NMe)_3$); [1]H ($\delta$ ppm) ($CDCl_3$): 0~0.3 (m, $Si(CH_3)_3$); 2~3.5 (m, $NCH_3$); 4.8~5.2 (m, SiH); [29]Si ($\delta$ ppm) ($CDCl_3$): -20.3 (m, $HSiN_3$); 1.3 (m, $NSi(CH_3)_3$); FT-IR (KBr pellet, $cm^{-1}$): 2955, 2896 ($\nu CH_3$); 2812, 1070 ($\nu NCH_3$); 2199 ($\nu SiH$); 1478, 1382, 1340 ($\nu BN$); 911 ($\nu SiN$); 1251, 837, 758 ($\nu SiC$).

**Oxygen contents were found to be < 2 wt% and therefore omitted. The composition was recalculated by reference to 100%.

*** Anal. found. (wt %) Si: 27.94; B: 8.98; N: 39.12; C: 23.96.

REFERENCES

[1]P. Miele, S. Bernard, D. Cornu, and B. Toury, Recent developments in polymer-derived ceramic fibers (PDCFs): Preparation, properties and applications - A review, Soft Materials, 4, 249-86 (2006).

[2]S. Bernard, M. Weinmann, P. Gerstel, P. Miele, and F. Aldinger, Boron-modified polysilazane as a novel single-source precursor for SiBCN ceramic fibers: Synthesis, melt-spinning, curing and ceramic conversion, J Mater Chem, 15, 289-99 (2005).

[3]H. P. Baldus, M. Jansen, and D. Sporn, Ceramic fibers for matrix composites in high-temperature engine applications, Science, 285, 699-703 (1999).

[4]R. Riedel, A. Kienzle, W. Dressler, L. Ruwisch, J. Bill, and F. Aldinger, A siliconboron carbonitride ceramic stable to 2000 °C, Nature, 382, 796-98 (1996).

[5]M. Takamizawa, T. Kobayashi, A. Hayashida, and Y. Takeda Method for the preparation of an inorganic fiber containing silicon, carbon, boron and nitrogen, US 4 604 367, 1986.

[6]T. Wideman, E. Cortez, E. E. Remsen, G. A. Zank, P. J. Carroll, and L. G. Sneddon, Reactions of monofunctional boranes with hydridopolysilazane: synthesis, characterization, and ceramic conversion reactions of new processible precursors to SiNCB ceramic materials, Chem Mater, 9, 2218-30 (1997).

[7]T. Wideman, K. SU, E. E. Remsen, G. A. Zank, and L. G. Sneddon, Synthesis, characterization, and ceramic conversion reactions of borazine/silazane copolymers: new polymeric precursors to SiNCB ceramics, Chem Mater, 7, 2203-12 (1995).

[8]K. Su, E. E. Remsen, G. A. Zank, and L. G. Sneddon, Synthesis, characterization, and ceramic conversion reaction of borazine-modified hydridopolysilazane: new polymeric precursors to SiNCB ceramic composites *Chem Mater*, **5**, 547-56 (1993).

[9]H.-P. Baldus, N. Perchenek, A. Thierauf, R. Herborn, and D. Sporn Ceramic fibers in the system silicon-boron-nitrogen-carbon, US 5 968 859, 1999.

[10]M. Jansen, B. Jaschke, and T. Jaschke, Amorphous multinary ceramics in the Si-B-N-C system. In *High Performance Non-Oxide Ceramics I*, Jansen, M., Ed. Springer-Verlag: Berlin, Heidelberg, 2002; Vol. 101, pp 137-91.

[11]M. Jansen, U. Muller, J. Clade, and D. Sporn Silicoboroncarbonitrogen ceramics and precurosr compounds, method for the production and use thereof, US 7 297 649 B2, 2007.

[12]S. Bernard, M. Weinmann, D. Cornu, P. Miele, and F. Aldinger, Preparation of high-temperature stable Si-B-C-N fibers from tailored single source polyborosilazanes, *J Eur Ceram Soc*, **25**, 251-56 (2005).

[13]M. Weinmann, M. Kroschel, T. Jaschke, J. Nuss, M. Jansen, G. Kolios, A. Morillo, C. Tellaecheb, and U. Nieken, Towards continuous processes for the synthesis of precursors of amorphous Si/B/N/C ceramics, *J Mater Chem*, **18**, 1810-18 (2008).

[14]J. Lee, D. P. Butt, R. H. Baney, C. R. Bowers, and J. S. Tulenko, Synthesis and pyrolysis of novel polysilazane to SiBCN ceramic, *J Non-cryst Solids*, **351**, 2995-3005 (2005).

[15]Y. Tang, J. Wang, X.-D. Li, H. Wang, W.-H. Li, and X.-Z. Wang, Preceramic polymer for Si-B-N-C fiber via one-step condensation of silane, BCl3, and silazane, *J Appl Polym Sci*, **110**, 921-28 (2008).

[16]D. F. Shriver, and M. A. Drezdz, *The manipulation of Air-Sensitive Compounds*. 2nd ed.; Wiley: New York, 1986.

[17]B. Toury, S. Bernard, D. Cornu, F. Chassagneux, J.-M. Letoffe, and P. Miele, High-performance boron nitride fibers obtained from asymmetric alkylaminoborazine, *J Mater Chem*, **13**, 274-79 (2003).

[18]B. Toury, and P. Miele, A new polyborazine-based route to boron nitride fibres, *J Mater Chem*, **14**, 2609-11 (2004).

[19]J. Haberecht, R. Nesper, and H. Grutzmacher, A construction kit for Si-B-C-N ceramic materials based on borazine precursors, *Chem Mater*, **17**, 2340-47 (2005).

[20]T. Jaschke, and M. Jansen, A new borazine-type single source precursor for Si/B/N/C ceramics, *J Mater Chem*, **16**, 2792-99 (2006).

[21]P. Miele, B. Toury, D. Cornu, and S. Bernard, Borylborazines as new precursors for boron nitride fibres, *J Organomet Chem*, **690**, 2809-14 (2005).

[22]T. Wideman, E. E. Remsen, E. Cortez, V. L. Chlanda, and L. G. Sneddon, Amine-Modified Polyborazylenes: Second-Generation Precursors to Boron Nitride, *Chem Mater*, **10**, 412-21 (1998).

PREPARATION OF SIC CERAMIC FIBERS CONTAINING CNT

Xiaodong Li, Haizhe Wang, Dafang Zhao and Qingling Fang

State Key Lab of Ceramic Fibers & the Composite
National University of Defense Technology
Changsha, 410073, China

ABSTRACT

SiC/CNT composite fibers were prepared by blending pretreated CNT (0.1-1 wt%) into polycarborsilane (PCS) followed by melt-spinning, curing and pyrolysis. The properties of the fiber may be significantly improved since the CNT's are mostly oriented along the fiber axis.

1. INTRODUCTION

SiC fiber is now widely applied in high-tech due to its desirable properties such as oxidation resistance, high hardness, low thermal expansion and high chemical stability[1]. Up to now, polymer-derived SiC fibers have been extensively studied since Yajima's synthesis[2] of polycarbosilane (PCS) as precursor. In order to improve the properties of polymer-derived SiC fibers, in this work, the preparation of SiC

fibers containing CNT are studied. Because of its unique structural[3,4] , mechanical and electronic

properties[5,6], CNT is promised to play an important role in reinforcing the ceramic fiber for both structural and functional applications.

2. EXPERIMENTAL PROCEDURE

PCS (melting point 210°C, molecular weight 2000, C/Si ratio 1.6) was dissolved in xylene. CNT,

prepared by CVD and pretreated with vinyl functional group, was added into the solution (the weight ratio of CNT:PCS of 0.1-1%) with agitation and ultrasonic dispersion. The dispersed solution was distilled under $N_2$ atmosphere and then in vacuum to remove the solvent. PCS/CNT precursor was

melt-spun and then cured in air up to 210°C followed by pyrolysis at 1200°C in $N_2$ atmosphere. Black

and metal-lustrous SiC/CNT fiber was thus obtained.

3. RESULT AND DISCUSSION

Since the presentation of CNT in PCS, it was found that the melting point of the blended

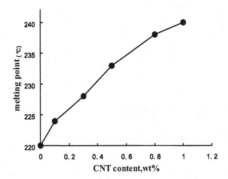

Fig.1 Relationship between melting point and CNT content of the blended precusor precursor was increased.

The distribution of CNT in PCS is very important for spinning. Fig.2 shows that the

Fig.2 The SEM of PCS/CNT precursor (JEOL JSM-5600LV)

pretreated CNT is distributing uniformly in the PCS precursor without any entanglement and aggregation.

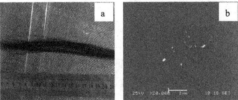

Fig.3 The PCS/CNT fiber (a) Image of spun fibers; (b) Cross section SEM of the fiber

The precursor thus formed can be melt-spun into flexible fibers, shown in Fig. 3(a), with diameters about 25μm. It is interesting to find that, in Fig. 3(b), the CNT's show points in the cross-section of the fiber. By comparison with Fig. 2, that means the CNT's are mostly oriented along the axis of the fibers owing to the extrusion and drawing in the spinning.

Fig.4 The SEM of SiC/CNT fiber (a) Surface; (b) Cross section

After air-curing and pyrolysis, PCS/CNT fiber turned into flexible SiC/CNT fiber, shown in Fig. 4(a), with smooth surface and diameter in the range of 20-25μm. The cross-section SEM image of the broken SiC/CNT fiber Fig. 4(b) shows that some CNT was pull out from the fiber.

4. CONCLUSIONS

1) CNT can be blended in PCS by ultrasonic dispersion followed by melt-spinning, air-curing and pyrolysis to prepare SiC/CNT fiber.
2) CNT is mostly oriented along the axis of the fibers, that may improve the properties of the ceramic fibers.

REFERENCES

[1] Russell, J. D., *High-performance synthesis fibers for composites.* . National Academy Press: Washington D. C., 1992; p 1-129.

[2]Yajima.S,Hayashi.J,Omori.M.Chem.Lett.1975,931.

[3]Copper,C.A,Young.R.J,Halsall.M.Composites,part A 2001,32A,401-411

[4]Gao.G,Cagin.T,Goddard.W.A,Nanotechnology 1998,9,184-191

[5]Uchida.T,Kumar.S.J.Appl.Polym.Sci.2005,98,985-989

[6]De Heer Walt,A .*MRS Bull.2004,29,281-285*

# PREPARATION AND PROPERTIES OF NON-CIRCULAR CROSS-SECTION SiC FIBERS FROM A PRECERAMIC POLYMER

Wang Yingde, Liu Xuguang, Wanglei, Lan Xinyan, Xue Jingen, Jiang Yonggang, Zhong Wenli

State Key Lab of Advanced Ceramic Fibres and Composites
College of Aerospace and Materials Engineering, National University of Defense Technology
Changsha, Hunan, China

## ABSTRACT

Preceramic polymer SiC fibres are usually circular in cross section. Polycarbosilane(PCS)-based fibres are melt spun and extruded into a wide variety of non-circular shapes. The trilobal, C-shaped, hollow, ribbonlike, pentagonal and swirl-shaped PCS fibres were prepared. These non-circular fibres were cured and pyrolyzed at the same conditions as the circular SiC fibres and the tensile strength, microstructure, electromagnetic properties and reverberation attenuation to microwave of the non-circular fibres were compared to that of circular SiC fibres. Results showed that non-circular SiC fibres have a higher tensile strength than that of the circular SiC fibres and exhibit an excellent microwave-absorbing property.

## INTRODUCTION

SiC fibres, which have many attractive properties such as high tensile strength and high elastic modulus, high creep and oxidation resistance, are promising candidates for reinforcing ceramic matrix composites. The fibres can be obtained by controlled pyrolysis of Polycarbosilane(PCS), which were melt spun to PCS green fibres, then the green fibres were cured in air and pyrolyzed in $N_2$ atmosphere under tension[1-4].

SiC fibres prepared by pyrolyzed process is a kind of n-semiconductor which conductance can be adjusted. Then SiC fibres could have some property in absorbing microwave. In X-band, the electromagnetic parameters of normal circular SiC fibres are $\varepsilon'$=3~5, $\varepsilon''$=0, $\mu'$=0.98~1.03, $\mu''$=0, resistance is about $106\Omega\cdot$cm[5], electromagnetic attenuation is little, and microwave can permeate it. The electromagnetic parameters and resistance of the SiC fibres can be adjusted by adjusting the composition of the precursor or changing the pyrolysis process. When the resistance is in the range of $101$~$103$ $\Omega\cdot$cm, SiC fibres have the best microwave-absorbing property in X-band[6].

Polycarbosilane(PCS)-derived SiC fibres are usually circular in cross section. PCS-based fibres are melt spun and extruded into a wide variety of non-circular shapes by using special spinning jet. It was found[7,8]that SiC fibres with trilobal and C-shaped cross section exhibit a particular property compared with the circular SiC fibres.

Microwave-absorbing structural composites (MASC) are important as a type of functional material. Non-circular SiC fibres, which have many attractive properties such as low density, high strength, high module, excellent high temperature resistance and appropriate microwave-electromagnetic parameters, are promising candidates for reinforcing MASC.

Based on the previous work, in order to enhance the research of non-circular SiC fibres, in the paper, trilobal, C-shaped, hollow, ribbonlike, pentagonal and swirl-shaped PCS fibres were melt spun. These non-circular fibres were cured and pyrolyzed at the same conditions as the circular SiC fibres. Then the tensile strength, microstructure, microwave-electromagnetic properties and reverberation attenuation of the non-circular SiC fibres were compared to that of circular fibres.

EXPERIMENTAL PROCEDURES

Polycarbosilane (PCS), which is a transparent yellow solid with a number average molecular weight of about 1500, was synthesized. PCS green fibres were prepared by melt-spinning equipments. Several kinds of special spinneret were used and six kinds of fibres including trilobal, C-shaped, hollow, ribbonlike, pentagonal and swirl-shaped fibres were prepared. These non-circular fibres were cured and pyrolyzed at the same conditions as the circular SiC fibres.

The tensile strength, microstructure, microwave-electromagnetic properties and reverberation attenuation of the non-circular fibres were compared to that of circular fibres. The tensile strength of single fibre was measured by the ergometer (YG001A, China). The microstructure of fibres was investigated by XRD(D5000, Siemens, Germany). The cross-sections were observed by SEM (X-650, Japan). And the electromagnetic properties were evaluated by scanning double six-port network analyzer. Specimens for the electromagnetic testing is 22.86×10.16×(2.0±0.1)mm. The reverberation attenuation was measured by network analyzer with RAM arcuate method (8720ET, Agilent).

RESULTS AND DISCUSSIONS

1 Microstructure of non-circular SiC fibres

Fig.1 shows SEM photographs of the non-circular SiC fibres, including the trilobal, C-shaped, hollow, ribbonlike, pentagonal and swirl-shaped fibres. The properties of non-circular SiC fibres are different for their diversification in cross-sections.

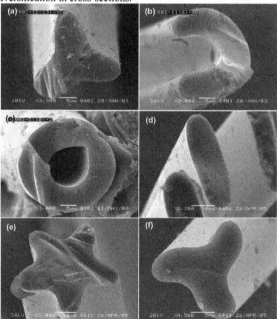

Fig.1 SEM photographs of the non-circular SiC fibres
(a: Trilobal, b: C-shaped, c: Hollow, d: Ribbonlike, e: Pentagonal, f: Swirl-shaped)

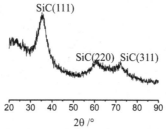

Fig.2 XRD pattern of non-circular SiC fibres

Fig.2 is the XRD pattern of the non-curcular SiC fibres. It is illustrated that non-circular SiC fibres were mainly made up of amorphous SiC and β-SiC crystallite, which is similar to the circular SiC fibres[9]. Consequently, it is concluded that the differences between circular and non-circular fibres is their the changes of cross-section.

2 Tensile strength of non-circular SiC fibres

The tensile strength of the non-circular SiC fibres are shown in Fig.3. The tensile strength of trilobal fibres is 1.5-2.1GPa, effective filament diameters(Ed) is 14-24μm, correspondingly, C-shaped is 1.2-1.7GPa and 22-28μm, hollow fibres is 1.3-1.8GPa with the Ed of 14-25μm.

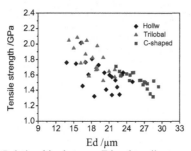

Fig.3 Relationships between Ed and tensile strength of
hollow, trilobal and C-shaped SiC fibres

The tensile strength of trilobal fibres is higher than that of C-shaped and hollow fibres, and C-shaped is almost equal to hollow one.

From Fig.4, the tensile strength of robbonlike fibres is 1.1-2.4GPa with the Ed of 21-32μm, the pentagonal fibres is 0.7-2.0GPa and 22-35μm, swirl-shaped is 1.0-2.4GPa and 23-34μm. Robbonlike fibres' strength is almost equal to swirl-shaped ones, and these two typed fibres are higher than pentagonal fibres.

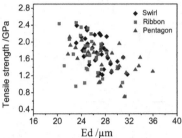

Fig.4 Relationships between Ed and tensile strength of
robbonlike, pentagonal and swirl-shaped fibres

3 Electromagnetic properties of the non-circular SiC fibres

Generally, $\mu'$ of the non-circular SiC fibres is 1 in X-band (8.2-12.4GHz), $\mu''$ is 0, which is similar to the circular SiC fibres. And in the paper, electromagnetic properties of non-circular SiC fibres are investigated in the same way.

From Fig.5 and Fig.6, $\varepsilon'$ of trilobal SiC fibres is between 6.68 and 6.82, $\varepsilon''$ is between 3.06 and 4.03. According to $\varepsilon'$ and $\varepsilon''$, tan$\delta$ of trilobal fibres can be calculated (tan$\delta$=0.46~0.59), which is illustrated in Fig.8.

Tangent of the loss angle of non-circular SiC fibres can be described by tan$\delta$. According to Fig.7, it can be known that the trilobal SiC fibre is a kind of dielectric attenuation material.

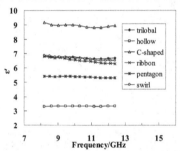

Fig.5 Relationships between $\varepsilon'$ and frequency of
electromagnetic wave of non-circular SiC fibres

As is shown in Fig.5 and Fig.6, $\varepsilon'$ of C-shaped SiC fibres is between 8.7 and 9.1, $\varepsilon''$ is between 4.7 and 5.9. And tan$\delta$ of C-shaped fibres is in the range of 0.54 ~ 0.65. So the C-shaped SiC fibre is also a kind of dielectric attenuation material.

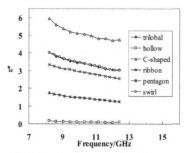

Fig.6 Relationships between ε″ and frequency of
electromagnetic wave of non-circular SiC fibres

ε′ of hollow SiC fibres is between 3.2 and 3.3, ε″ is between 0.1 and 0.2. According to ε′ and ε″, tanδ of hollow fibres is between 0.03 and 0.05 (see Fig.7), and the hollow SiC fibre is also a kind of dielectric material. ε′ of ribbonlike SiC fibres is between 6.2 and 6.8, ε″ is between 2.5 and 3.4. And tanδ of ribbonlike fibres is between 0.40 and 0.50. The ribbonlike SiC fibre also is a kind of dielectric attenuation material.

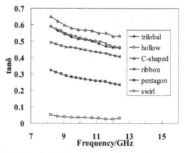

Fig.7 Relationships between tanδ and frequency of
electromagnetic wave of non-circular SiC fibres

ε′ of pentagonal SiC fibres is between 5.2 and 5.4, ε″ is between 1.2 and 1.8. From Fig.8, tanδ of pentagonal fibres is between 0.23 and 0.33. The pentagonal SiC fibre is a kind of dielectric attenuation material too. Like the presentations of SiC fibres above, ε′ of swirl-shaped SiC fibres is between 6.5 and 6.9, ε″ is between 3.0 and 4.0. And tanδ of swirl-shaped fibres is between 0.46 and 0.58. The swirl-shaped SiC fibre is also a kind of dielectric attenuation material.

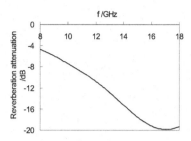

Fig.8 The reverberation attenuation of composite prepared by trilobal SiC fibres

Reverberation attenuation of trilobal SiC fibres was evaluated (Fig.8). It is shown that the largest attenuation is 19.6dB, the frequency range with attenuation above 10dB is between 11 and 18GHz. According to the analysis of non-circular fibres' electromagnetic parameters and the reverberation attenuation of trilobal SiC fibres, it can be conclude that trilobal, C-shaped, ribbonlike and swirl-shaped SiC fibres have good microwave-absorbing properties because of their large medium electrical wastes. Hollow and pentagonal fibres also have some microwave-absorbing properties.

CONCLUSIONS

(1) Non-circular SiC fibres were made up of amorphous SiC and $\beta$-SiC crystallite.

(2) Compared with circular SiC fibres with the same Ed, the tensile strengths of non-circular SiC fibres increase with 10-20%.

(3) In X-band, $\mu'$ of non-circular SiC fibres is 1, $\mu''$ is 0. $\varepsilon'$ of trilobal SiC fibres is between 10.1 and 13.9, $\varepsilon''$ is between 7.9 and 8.9.

(4) In X-band, trilobal, C-shaped, ribbonlike and swirl-shaped SiC fibres have good microwave-absorbing properties, the largest attenuation is 19.6dB, the frequency range with attenuation above 10dB is between 11 and 18GHz.

REFERENCES

[1] S. Yajima, J. Hayashi, M. Omori, Continuous silicon carbide fibre of tensile strength, *Chem.Lett.*, **9**, 931-934(1975).

[2] S.Yajima, J. Hayashi, M. Omori, et al, Development of silicon carbide fibres with high tensile strength, *Chem. Lett.*, **2**, 683-685(1976).

[3] D. B. Fishichbach, P. M. Lemoine, Mechanical properties and structure of a new commercial SiC-type fibre Tyrranno, *J. Mater. Sci.*, **23**, 987-993(1988).

[4] K. Kumagawa, Fabrication and mechanical properties of new improved Si-M-C-(O) Tyrrano fibre, *Ceram. Eng. Sci. Proc.*, **19**(1), 65-72(1998).

[5] N. Muto, M. Miyayama, Yahagida, Infrared Detection by Si-Ti-C-O Fibres, *J. Amer.Ceram.Soc.*,**73**, 443-450(1990).

[6] T. Yamauna, T. Toshikawa, M. Shibuya, *Electromagnetic wave absorbing material*, USP5094907(1992).

[7] Y. D. Wang, Y. M. Chen, M. F. Zhu, et al., Development for silicon carbide fibres with trilobal cross section, *J. Mater. Sci. Lett.*, **21**, 349-350(2002).

[8] Y. G. Jiang, Y. D. Wang, X. Y. Lan, et al, Preparation of C-shaped silicon carbide fibres, *J. Mater. Sci. Lett.*, **39**, 5881-5882(2004).

[9] M. Takeda, J. Sakamoto, Y. Imai, et al, Thermal stability of the low-oxygen-content silicon carbide fibre Hi-Nicalon[TM] , *Comp. Sci. Technol.*,**59,** 813-819(1999).

ECONOMY OF FUEL GAS IN A COMBUSTION FURNACE BY MEANS OF Si-C-Zr-O TYRANNO-FIBER MAT SHEETS CONVERTING HIGH TEMPERATURE GAS ENTHALPY INTO RADIANT HEAT RAYS

Suzuki, Kenji[1]; Ito, Kiyotaka[2]; Tabuchi, Matsumi[3]; Shibuya, Masaki[4]

[1] Advanced Institute of Materials Science, Moniwadai 2-6-8, Taihaku-ku, Sendai 982-0252, Japan

[2] Asahi Seisakusho Co., Ltd., Saitama 339-0078, Japan

[3] Niigata Furnace Kogyo Co., Ltd., Niigata 950-0801, Japan

[4] Ube Industries, Ltd., Ube 755-8633, Japan

ABSTRACT

Si-C-Zr-O Tyranno fiber mat sheet efficiently works as a thermo-filter for collecting high temperature exhaust gas enthalpy, which is converted into infrared heat rays radiating toward the inside of a gas combustion furnace. When a thermo-filter of the Si-C-Zr-O Tyranno fiber mat sheet of 6 mm thick and 95 % opening space is fixed at the exit of 1300 K exhaust gas in a propane gas fuel furnace, saving of the fuel gas to be spent reaches more than 20 % compared with no thermo-filter. Additional 10 % saving of the fuel gas is obtained by pasting the Si-C-Zr-O Tyranno fiber mat sheet as a thermo-reflector on the walls inside the furnace. Furthermore, the Si-C-Zr-O Tyranno fiber mat sheet thermo-filter and –reflector totally accelerate heating rate and emphasize uniform temperature distribution in the furnace.

INTRODUCTION

The most remarkable thermal loss in a gas combustion furnace is usually originated from high temperature exhaust gas. Therefore, an important key-point for saving fuel gas depends on how much the enthalpy of high temperature exhaust gas can be recovered. Recovery of the exhaust gas enthalpy has been often achieved by producing high temperature-pressure steam to operate turbine power generators, and/or by preheating fuel gas and air before the exhaust gas is released out from a chimney. However, these conventional ways for recovering the exhaust gas enthalpy need a complicated mechanical and electric system to lead into high cost performance in construction and operation.

If the high temperature exhaust gas enthalpy can be directly recovered into a furnace, consumption of fuel gas must be reduced without high cost recovering system. Such an idea has been so far realized by pasting heat radiant materials on the wall of a furnace or by spraying radiant particles emitting infrared rays in blazing flame. In particular, Echigo[1] has investigated theoretically the mechanism of recovering directly the gas enthalpy by the porous metallic filter which is equipped in the path of exhaust gas in a chimney.

In this paper, it has been demonstrated that Si-C-Zr-O Tyranno fiber[2] mat sheet works well as a thermo-filter for collecting high temperature exhaust gas enthalpy to convert into infrared rays heating back the inside of a gas combustion furnace. By using the Si-C-Zr-O Tyranno fiber mat sheet as a thermo-filter of optimized free space and thickness, saving of fuel gas to be spent reaches more than 20 % compared with the case of no thermo-filter. Additional 10 % saving of fuel gas is obtained by pasting the Si-C-Zr-O Tyranno fiber mat sheet as a thermo-reflector on the walls inside a gas combustion furnace. Simultaneous use of both the thermo-filter and –reflector provides a characteristic advantage for increase in heating rate and uniform temperature distribution in the furnace.

STRUCTURE AND PROPERTIES OF Si-C-Zr-O TYRANNO FIBER MAT SHEET
The structure[3,4] and properties[2] of the Si-C-Zr-O Tyranno fiber mat sheet to be used as a thermo-filter and-reflector are characterized as follows:
(1) The chemical composition of Si-C-Zr-O Tyranno fiber used is 56.6wt%Si, 34.8wt%C, 1.0wt%Zr and 7.6wt%O, and the C/Si atomic ratio is 1.44.
(2) X-ray diffraction and small angle X-ray scattering show that the Si-C-Zr-O Tyranno fiber has an amorphous matrix phase, in which $\beta$-SiC nano-crystalline particles surrounded by $sp^2$-C nano-layer precipitate together with nano-voids but nano-clusters of $SiO_2$ glass are not found[5].
(3) The Si-C-Zr-O fiber mat sheet has a very large opening space of 90~95% to provide only a few $mmH_2O$ pressure loss for gas to penetrate the mat sheet of a 6 mm in thickness.
(4) The Si-C-Zr-O fiber has a thin diameter of about 10 μm, on whose surface $SiO_2$ glass thin layer is formed to protect the fiber from further oxidation even in high temperature air atmosphere.   .
(5) The effective contact surface between fiber and gas in the mat sheet is quite large (3000 $cm^2$/g) to result in an extraordinarily large thermal transfer coefficient of the mat sheet.
(6) The Si-C-Zr-O fiber shows a considerably low specific heat (0.71 J/gK) and thermal conductivity (2.52 W/mK).
(7) The Si-C-Zr-O fiber does efficiently irradiate infrared heat rays at high temperatures above 1000 K like an ideal black body does ($\sim(\Delta T)^4$).

Currently it has been often reported that some thermal properties are much emphasized in nano-fluids[6]. Therefore, the characteristic behaviors of the Si-C-Zr-O Tyranno fiber mat sheet mentioned above may come from the mechanism of an assembly of multi-nano-phases similar to nano-fluids.

RECOVERY OF HIGH TEMPERATURE EXHAUST GAS ENTHALPY BY Si-C-Zr-O TYRANNO FIBER MAT SHEET

Thermo-filter effect
When high temperature exhaust gas passes through Si-C-Zr-O fiber mat sheet, the enthalpy of the exhaust gas is efficiently filtered off by the mat sheet. Therefore, the temperature of the exhaust gas is drastically reduced.   On the other hand, the gas enthalpy filtered is converted on the mat sheet into

infrared heat rays to radiate toward upper side of exhaust gas flow, that is exactly the inside of the furnace. This means that fuel gas to be spent is saved by the amount of infrared radiation heating and the heating rate inside the furnace is accelerated with more uniform temperature distribution. This scenario has been confirmed by Echigo's thermo-engineering calculation[1].

In order to confirm the characteristic behaviors of the Si-C-Zr-O fiber mat sheet as a thermo-filter, we observed preliminarily the temperature difference of the exhaust gas at the both surfaces of the mat sheet before and after passing through the Si-C-Zr-O fiber mat sheet thermo-filter. The arrangement of the measurement is shown in Fig. 1. A small butane gas burner of 1000 kcal/h was located at the bottom of a ceramic cylinder of 100 mm in inner diameter and 200 mm in length, and then the Si-C-Zr-O Tyranno fiber mat sheet (Ube Industries Ltd., ZF-S250B) of 6 mm in thickness was fixed on the top of the ceramic cylinder. The temperatures of the exhaust gas at the both surfaces of the mat sheet were measured by thermo-couples fixed on the each surface of the mat sheet, as shown in Fig. 1. The results are shown in Fig. 2. The temperature gap ($\Delta T = T_i - T_o$) between the inside surface ($T_i$=800 K; blue line 1) and outside surface ($T_o$=600 K; purple line 3) is $\Delta T$ =200 K on the Si-C-Zr-O Tyranno fiber mat sheet, while the temperature gap is only $\Delta T$ =40 K without Si-C-Zr-O Tyranno fiber mat sheet.

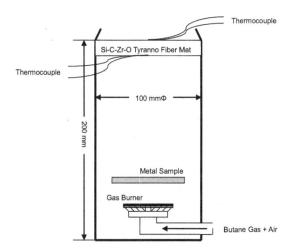

Figure 1. Experimental arrangement for exploring thermo-filter effect of Si-C-Zr-O Tyranno fiber sheet mat (ZF-S250B ; Ube Industries Ltd.).

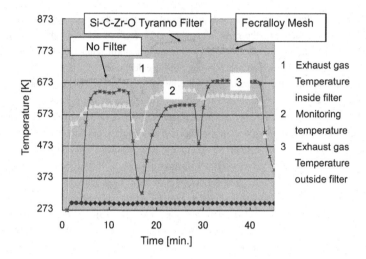

Figure 2. Temperature gaps of exhaust gas caused by thermo-filters of Si-C-Zr-O Tyranno-fiber mat sheet (ZF-S250B; Ube Industries Ltd.) and Fecralloy wire mesh.

The recovery ratio of the gas enthalpy in the preliminary experiment can be roughly estimated as $(\Delta T/Ti) \times 100 = (200/800) \times 100 = 25\%$ for the Si-C-Zr-O Tyranno fiber mat sheet. For comparison, we have measured the temperature gap of the exhaust gas passing through a Fecralloy wire mesh sheet (Fe-Cr-Al-Si alloy, 0.4 mm diameter wire x 16 mesh). The recovery ratio for the Fecralloy wire mesh sheet is about 13 %. Since the infrared heat rays are emitted less and heat loss from Fecralloy is more than those from the Si-C-Zr-O Tyranno fiber, the net recovery ratio must be much less than 13 %. Furthermore, the Si-C-Zr-O Tyranno fiber mat sheet is more promising as a thermo-filter in air because of superior anti-oxidation in the high temperature above 1300 K.

Thermo-reflector effect
    When high temperature exhaust gas touches the surface of the Si-C-Zr-O Tyranno fiber mat sheet pasted on thermo-insulation walls in a gas combustion furnace, a part of the gas enthalpy is converted to infrared heat rays radiating into the inside of the furnace. This is a thermo-reflector effect. Since opening space in the Si-C-Zr-O Tyranno fiber mat sheet is finely and uniformly distributed, the exhaust gas does not only touch the mat sheet but penetrate slightly into the mat sheet, Therefore, the

thermo-reflector effect is much enhanced in the Si-C-Zr-O Tyranno fiber mat sheet, compared with a solid insulation board. We can expect higher cooling rate because of a very high emission rate of heat rays from the mat sheet when the door of the furnace is opened.

ECONOMY OF FUEL GAS IN A COMBUSTION FURNACE BY Si-C-Zr-O TYRANNO FIBER MAT SHEETS

We have verified how much fuel gas and heating time are saved in a propane gas combustion furnace in practical use and then in actual operation of heat treatment of cast iron automobile engine parts by using the Si-C-Zr-O Tyranno fiber mat sheet thermo-filter and –reflector.

Verification in a practical use propane-gas combustion furnace

The specification of a propane-gas combustion furnace used for verifying test is summarized in Table 1. The Si-C-Zr-O Tyranno fiber mat sheet thermo-filter was fixed on the exit of exhaust gas in the furnace. As shown in Fig. 3, the Si-C-Zr-O Tyranno fiber mat sheet thermo-reflector was pasted on the walls and ceiling in the furnace. Temperatures in the furnace were monitored as a function of burning time by several thermo-couples installed on and in the iron blocks (65mm x 65mm x 220mm) settled on the bottom floor of the furnace. The temperatures of exhaust gas at the inner and outer surface of the mat sheet thermo-filter and consuming rate of propane fuel gas were measured as a function of burning time. For comparison, the same measurements were repeated for the cases with and without the Si-C-Zr-O Tyranno fiber mat sheet thermo-filter and reflector. Fig. 4 is the temperature changes for no filter and no reflector, while Fig. 5 shows the temperature changes for the case of only the filter. Results

Table 1. Specification of a gas combustion furnace used for verifying fuel economy
by Si-C-Zr-O fiber mat sheet thermo-filter and –reflector

| Size of combustion room | 900 mm wide x 830 mm deep x 1300 mm high |
|---|---|
| Thermal insulation | Wall : $Al_2O_3$-$SiO_2$ fiber mat of 200 mm thick<br>Floor : Firebrick of 300 mm thick |
| Power rate of combustion burner | 100,000 kcal/h |
| Fuel gas | Propane (Combustion heat : 22,450 kcal/$m^3_N$) |
| Air-to-fuel gas ratio | 1.2 |
| System of exhausting gas | Natural convection by a chimney of 300 mm in diameter |
| Pressure loss in combustion furnace | 10 mm $H_2O$ |

Figure 3. Inside of propane gas combustion furnace equipped with Si-C-Zr-O Tyranno fiber
sheet mat thermo-filter and -reflector

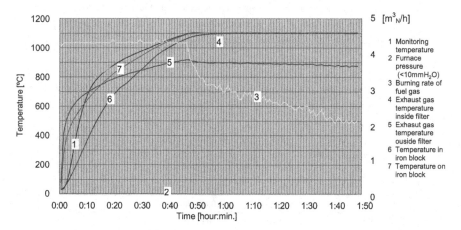

Figure 4. Temperature variation and propane gas burning rate during combustion experiment
of the furnace without Si-C-Zr-O Tyranno fiber mat sheet thermo-filter and –reflector.

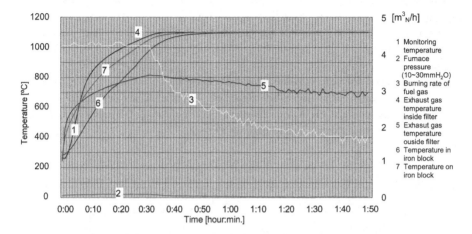

Figure 5. Temperature variation and propane gas burning rate during combustion experiment of the furnace with Si-C-Zr-O Tyranno fiber mat sheet thermo-filter but no thermo–reflector.

Table 2. Results of combustion experiment for verifying fuel economy
by Si-C-Zr-O Tyranno fiber sheet mat thermo-filter and –reflector

| Experiment No. | 1 | 2 | 3 | 4 |
|---|---|---|---|---|
| Thermo-filter | No | Yes | Yes | No |
| Thermo-reflector | No | Yes | No | Yes |
| Duration of reaching 1373 K in furnace (min.) | 48 | 29 | 30 | 41 |
| Exhaust gas temperatures (K) at inner side of filter (T1) | 1373 | 1373 | 1373 | 1373 |
| outer side of filter (T2) | 1343 | 873 | 973 | 1273 |
| gap ($\Delta T=T1-T2$) | 30 | 500 | 400 | 100 |
| Duration of reaching 1373 K in iron block (min.) | 61 | 48 | 53 | 60 |
| Amount of fuel gas spent ($m^3{}_N$) | 6.29 | 4.59 | 5.28 | 5.60 |
| Economy ratio of fuel gas (%)[#] | 0 | 27 | 16 | 11 |

#: based on case of no thermo-filter and no thermo-reflector

are summarized in Table 2 and show that the combination of thermo-filter and reflector leads to a huge amount of fuel economy reaching 30 % and the contribution of thermo-filter to thermo-reflector is about 2 to 1.

Fuel economy in heat treatment of cast iron automobile engine parts

Heat treatment of cast iron automobile engine parts was carried out in industrial scale of several tens tons per week by using the both thermo-filter and reflector of Si-C-Zr-O Tyranno fiber mat sheet. Table 3 shows that saving propane gas reached finally 39 %.

Table 3. Economy of fuel gas in heat treatment of cast iron automobile engine parts

|  | $1^{st}$ week | $2^{nd}$ week | $3^{rd}$ week |
|---|---|---|---|
| Thermo-filter | No | Yes | Yes |
| Thermo-reflector | No | Yes | Yes |
| Weight of cast iron automobile engine parts annealed (ton/week) | 54.9 | 53.4 | 55.1 |
| Volume of Propane gas spent ($m^3_N$/week) | 767.9 | 572.7 | 464.8 |
| Specific volume of propane gas used for annealing cast iron of 1 ton ($m^3_N$/ton) | 14.0 | 10.7 | 8.4 |
| Economy of fuel gas (%)[#] | 0 | 23.4 | 39.6 |

#: based on case of no thermo-filter and no thermo-reflector

CONCLUDING REMARKS

Based on the characteristic behavior of Si-C-Zr-O Tyranno fiber mat sheet, it was confirmed that high temperature exhaust gas enthalpy in a gas combustion furnace could be converted into radiant heat emitting infrared rays toward inside of the furnace by the Si-C-Zr-O Tyranno fiber mat thermo-filter and directly recovered in the furnace itself to result in huge economy of fuel gas to be spent. With the intimate combination of thermo-filter and thermo-reflector, fuel gas was saved by more than 30 % and heating rate is accelerated by about 40 % together with uniform temperature distribution in the furnace.

However, there is a difficult problem that each thin fiber of about 10 μm in diameter constructing the Si-C-Zr-O Tyranno fiber mat sheet receives drastically chemical and mechanical damages by strong and high speed blow of high temperature exhaust gas often reaching a velocity of 100 km/sec. Furthermore, the exhaust gas in the practical furnace in industrial use includes various kinds of fine particles of metal oxides and active chemical species to provide serious damages to the fiber. Since the Si-C-Zr-O Tyranno fibers used in this study have been produced by the Yajima process[7], it can not be neglected that the surface of the fiber is attacked in an ambient gas including steam at the temperature above 1500 K.

To overcome the difficulties mentioned above, it is necessary that the whole process of Si-C based fibers[8] has to be improved from the basic points including synthesis of new precursor polymers[9], spinning, organic-to-inorganic conversion and so on. Design of gas combustion furnaces also must be optimized for easy and safe equipment of the thermo-filter and –reflector.

REFERENCES

[1]R. Echigo, Effective Conversion Method between Gas and Thermal Radiation and Its Application to industrial Furnaces, *Transactions of Japan Society of Mechanical Engineers*, **B48**, 2315-2323(1982).

[2]K. Kumagawa, H. Yamaoka, M. Shibuya and T. Yamamura, Fabrication and Mechanical Properties of New Improved Si-M-C-(O) Tyranno Fiber, *Proc. 22nd Annual Conf. Composite, Advanced Ceramics, Materials and Structures A, Cocoa Beach, Florida (Edited by Bray, The American Ceramic Society)*, 65-72(1998).

[3]K. Suzuki, K. Kumagawa, T. Kamiyama and M. Shibuya, Characterization of the Medium-Range Structure of Si-Al-C-O, Si-Zr-C-O and Si-Al-C Tyranno Fibers by Small Angle X-ray Scattering, *J. Materials Science*, **37**, 949-953(2002).

[4]K. Suzuki, T. Kamiyama and M. Shibuya, Nano-Scale Composite Structure of Polymer-Route Si-C-M-O Fibers Characterized by Small-Angle X-Ray Scattering, *Key Engineering Materials*, **247**, 13-18(2003).

[5]K. Suzuya, S. Kohara, K. Okamura, H. Ichikawa and K. Suzuki, Intermediate-Range Order in Polymer Route Si-C-O Fibers by High-Energy X-Ray Diffraction and Reverse Monte Carlo Modelling, *to be published in the Proceedings of PACRIM-8* (Vancouver, 2009).

[6]J. Buongiorno, Center for Nanofluids Technology, Massachusetts Institute of Technologies, Boston, USA, http://web.mit.edu/nse/nanofluids/.

[7]S. Yajima, Special Heat-Resisting Materials from Organometallic Polymers, *American Ceramic Society Bulletin*, **62**, 893-903(1983).

[8]K. Okamura, T. Shimoo, K. Suzuya and K. Suzuki, SiC-Based Ceramic Fibers Prepared via Organic-to-Inorganic Conversion Process, *J. Ceramic Society of Japan*, **114**, 445-454(2006).

[9]S. Kyushin, H. Shiraiwa, M. Kubota, K. Negishi, K. Okamura and K. Suzuki, Poly[(silylyne)ethynylene] and Poly[(silylene)ethynylene] ; New Precursors for the Efficient Synthesis of Silicon Carbides, *to be published in the Proceedings of RIMPAC-8* (Vancouver, 2009).

# Author Index

Printed in the United States
By Bookmasters